Egypt Ignited

Egypt Ignited

How Fossil Capital Arrived on the Nile

Amr Khairy

VERSO
London • New York

First published by Verso 2026

The manufacturer's authorized representative in the EU for product safety (GPSR) is LOGOS EUROPE, 9 rue Nicolas Poussin, 17000, La Rochelle, France
contact@logoseurope.eu

1 3 5 7 9 10 8 6 4 2

Verso
UK: 6 Meard Street, London W1F 0EG
US: 207 East 32nd Street, New York, NY 10016
versobooks.com

Verso is the imprint of New Left Books

ISBN-13: 978-1-83674-056-8
ISBN-13: 978-1-83674-058-2 (US EBK)
ISBN-13: 978-1-83674-057-5 (UK EBK)

British Library Cataloguing in Publication Data
A catalogue record for this book is available from the British Library

Library of Congress Cataloging-in-Publication Data

Names: Khairy, Amr author
Title: Egypt ignited : how fossil capital arrived on the Nile / Amr Khairy.
Description: Brooklyn : Verso Books, 2026. | Includes bibliographical references and index.
Identifiers: LCCN 2026025005 (print) | LCCN 2026025006 (ebook) | ISBN 9781836740568 paperback | ISBN 9781836740582 ebk
Subjects: LCSH: Industrialization – Egypt – History – 19th century | Industrialization – Social aspects – Egypt | Industrialization – Economic aspects – Egypt | Fossil fuels – Egypt – History – 19th century | Capitalism – Egypt – History – 19th century | Egypt – History – 19th century
Classification: LCC DT100 .K53 2026 (print) | LCC DT100 (ebook) | DDC 962/.03 – dc23/eng/20260519
LC record available at https://lccn.loc.gov/2026025005
LC ebook record available at https://lccn.loc.gov/2026025006

Typeset in Minion by Hewer Text UK Ltd, Edinburgh
Printed and bound by CPI Group (UK) Ltd, Croydon, CR0 4YY

To Adam and Karma

Contents

Acknowledgements

While researching, writing, and refining this book between 2019 and 2026, I received crucial support that shaped and sustained my writing from mentors, friends, colleagues, and family. The manuscript evolved thanks to feedback and comments from Andreas Malm, Wim Carton, Alf Hornborg, Timothy Mitchell, Jenny Bulstrode, Martina Rieker, Aaron Jakes, Fred Albritton Jonsson, Marc Flandreau, and Mohamed Tawfiq.

Sections of the book were developed through the kind engagement of colleagues at the Department of Human Geography and Human Ecology Division at Lund University, between 2019 and 2023, where I received a doctoral degree, then at the French Institute for Oriental Archaeology (IFAO), and the Center for Economic, Legal, and Social Studies and Documentation (CEDEJ), the two institutions that hosted me as a post-doctoral fellow and made the completion of the book project possible, between 2024 and 2026. I am very grateful for the insights and support I received from Antoinette Ferrand, Frédéric Lagrange, Alya Elhosseiny, Mélanie Henry, Yasmin Shafie, Dina Bakhoum, Fatma Keshk, Pascal Menoret, Abbès Zouache, Mads Barbesgaard, Peter Svensson, Andreas Roos, Vasna Ramasar, Sanna Händén-Svensson, Mariko Takedomi-Karlsson, and the students of human ecology and on the CPS Masters programme at Lund University.

I also presented research for and sections of this book at the University of Cambridge during the 'Early Encounters with Coal' workshop organised by the Department of the History and Philosophy of Science; at the

NETWoRC seminar; at the Ottoman Political Economy workshop at Northumbria University; at the OPE online seminar series; at the Netherlands-Flemish Institute in Cairo; at the SOAN-SEA lecture series at the American University in Cairo; at the Italian Archaeological Center in Cairo (CAI); and at the Institute for Poverty, Land and Agrarian Studies (PLAAS) at the University of the Western Cape. These institutions and the collectives they hosted left me very grateful for feedback and support from Simon Schaffer, Richard Staley, Peter Hill, Ifdal El Saket, Dina Makram Ebaid, Hanan Sabea, Ruth Hall, and Jun Borras.

Then at Verso, Sebastian Budgen, Mark Martin, and Neil Burkey provided incredible editorial support that I truly perceive as magical.

Over the years, I received feedback and support from many other mentors, colleagues, and friends, including Ahmed Shokr, Hanan Hammad, Malak Rouchdy, Yasmine Hafez, Nermin Nizar, Abdallah Sobeih, Ayman Adam, Khaled Esmail, Essam Farghaly, Ihab Dakroury, Ralph Bodenstein, Abeer Rabei Abazeed, Saker El Nour, Eman Elnimr, Samuli Schielke, Amal Idris-Haroun, Osama Fekry, Santiago Gorostiza, and Hamza Hamouchene.

Finally, my family's limitless support sustained me and kept me grounded: my partner and wife Asmaa Askalany has been the kindest person in my life over the years. My daughter and son, Karma and Adam, were the ones I turned to for the most piercing and effective insights whenever the project hit an impasse. The unconditional love of my late mother, Wafaa Abdulhakim, and my siblings, Hatem, Haytham, and Osama, carried me effectively through moments of doubt, big and small.

Introduction

Things Ignited

Legend, on the tongues of peasants in Egypt in 1865, had it that a man named Ahmed Al-Tayeb started one night to recite softly one of the names of Allah. Three thousand times that night he summoned the kind god, '*Ya latif, ya latif, ya latif*', and he kept at this incantation every night for three years. Once he was done, at dawn, he rose invincible, the air around him buzzing with power, his flesh no longer vulnerable to the stabs of swords or the piercing of bullets. He left his home, somewhere in the village of Qau Gharb, at the Girga-Assiout line in Middle Egypt, and declared himself the Mahdi, he who comes at the end of time and kills the Antichrist. He called a revolution, and thousands of peasants followed. In the early spring of 1865, villages and hamlets woke to a morning teeming with angry masses, who attacked steamers and sailing cargo boats taking the agrarian commodities bought by foreign and local merchants across the Nile downstream to Cairo. They went after tax collectors and defied the state officials who came to collect corvée labourers. They confiscated money, mass-produced sugar, and grains. They brought gunpowder from the surrounding quarries in the hills at the far tips of the valley and they grabbed any gun they could find. An English woman in Luxor at the time was told by a friend that the rebels were seeking the radical redistribution of all wealth. The friend leaned close and looked at her fancy pocket watch. Speaking of the messianic *darweesh* Al-Tayeb, he said, 'He would break up your pretty clock and give every man a broken wheel out of it, and so with all things.'[1]

This almost totally forgotten uprising of Qau and its surroundings came at a critical moment of Egypt's integration, across the long decades of the precolonial nineteenth century, into Europe's industrial capitalism, and through steam technology purchased by loans, a process that brought colonialism. It was the same process that declared the arrival into Egypt of the Capitalocene, the epochal rise and entrenchment of fossil-fuelled economic activities that started the country's direct contribution to global warming through the burning of fossil fuels.

At the time, the Lancashire Cotton Famine, which had been caused by the American Civil War – in combination with other global and local economic, social, and ecological crises – led to a sudden and shocking chain of upheavals to the land and its people. Around Qau, the hotspot of the uprising, thousands of landless peasants and impoverished small landholders were forced to do corvée: in most cases, in big and small infrastructural projects to serve the cotton boom in Egypt that brought lucrative profits to Ismail Pasha (later Khedive Ismail, the semi-independent Viceroy to the Ottoman Sultan) and the richest landholders.

Just before they rose up, 50,000 men in the region were requisitioned to dig a 200-mile canal across Middle Egypt. The canal was needed to optimise the irrigation infrastructure that was to accommodate hundreds of steam pumps imported overnight from England. The Pasha had earlier recruited thousands more as soldiers, who were taken to the Upper Nile in Sudan to start cultivating cotton and to awaken the natives to the benefits of industriousness and civilisation. The regiments sent to Sudan revolted – along with the general populace – so the Pasha imposed a levy on the peasants of Middle Egypt to supply camels and food for a second batch of soldiers, sent into Sudan to execute the rebellious regiments and crucify their leaders. Soon after the Sudan affair, in Qau itself, the centre of the uprising, waves of troops disembarked from Nile steamers – no fewer than 10,000 soldiers, equipped with cannons, led by Ismail Pasha – to put an end to the rebellion and sweep the villages clean.

In a few days, a dozen villages were razed to the ground. Thousands fled to the desert. The Pasha sent paramilitary troops on camels in

1 For all relevant citations for this paragraph and the rest of this section, see Chapter 8.

pursuit, and thousands of men, women, children, and farm animals were killed or displaced. Fourteen hundred to 2,000 women, men, and children were killed, among them hundreds of men who were lined up by the dozen to be beheaded, and a few weeks later, the whole zone had become a banquet for dogs and vultures. The Mahdi was killed, some said, or had abandoned his physical form to fly away and on some other plane would bide his time to fight another day. His legend and the uprising were not, however, unique. Such events were common across Egypt's early modern history, especially in the decades following the 1820s, until the peasants articulated their anger around concepts of the nation, freedom, and justice in the Urabi Revolution that removed Khedive Ismail from office in 1879.

The Qau uprising is thus one moment among many across the decades that made up Egypt's precolonial nineteenth century – roughly the 1820s to the 1870s – and together form the socioecological phase of igniting Egypt, both its nature and its people. Across the world, fossil-fuelled technology was starting to stir outside England, the centre of the Industrial Revolution. Steam engines for the agrarian and industrial production of commodities were being sent, parts in crates across the ocean, to faraway places. They were assembled by European engineers here and there. Around the globe, the engine started to gather steam and send columns of smoke into the atmosphere. It was the ignition for a century-long process – of what we now recognise as global warming.

Egypt (among a few other places) was at the forefront of the first wave of this process of globalising fossil capital that ignited global warming. It was there, in the early 1820s, that one of the world's greatest iron foundries for manufacturing modern steam-powered spinning machines was established; the ruler of the country, Mehmed Ali Pasha, sent coal prospectors and miners with the troops that invaded Sudan and Syria across the 1820s and 1830s to bring back coal for his extensive factory machines, which he powered by oxen while he searched for coal. It was there, in the 1850s, that one of the first passenger railways anywhere in the world was built, and cotton ginning began to be steam-powered. It was there, in the early 1860s, that networks of fossil and financial capital, infused with industrial knowledge, running between London, Alexandria, and Cairo, extended steam power into agrarian production. It did this through the invention of the steam plough, deployed overnight in the Nile valley in numbers greater and processes more integrated than its counterparts in the English countryside, or indeed anywhere else in the world. It was there, at the turn of the 1870s, that alternative fuels – biomass – were experimented with inside fireboxes of engines and boilers to make up for the scarce coal. It was there, finally, that the first absurd and tragic accidents occurred outside Europe: steam engines poisoning labourers with carbon monoxide, or exploding and flying through the air only to burst through roofs hundreds of metres away on the way back down. It was in Egypt, a country we now categorise as part of the Global South, that industrial capitalism – the coupling of fossil-fuelled technologies with the mass production of commodities undergoing sustained, limitless growth – started to take hold somewhere outside of Europe.

This is Egypt ignited: the emergence of steam-powered technology on the Nile and the integration of that country into industrial capitalism. Whenever and wherever it went across the land, this economic engine was preceded and followed by crises of all sorts, prefiguring the disasters that would later become known as the hallmarks of this new epoch.

This Book

This book reconstructs historical processes that together formed the bulk of a packed half-century, 1820s–70s, during which Egypt was first exposed to steam power and associated technologies. The viewpoint from which this history comes together is the co-evolution over the decades of the means of production: the soil, animals – and humans when they, rather than animals, were compelled to power production machinery – and steam technology. The question from which I depart is: how did fossil-fuelled technology find its way to Egypt on its way to global dominance? This should not be mistaken for an exercise in teleology. The global dominance of such technology was not uncontested; to the contrary, the advances of the steam-powered complex were intensely contested, including in Egypt, where it provoked acts of resistance such as the Qau uprising. Egypt was not easily malleable. But we know that dominance was the ultimate outcome and must therefore ask how it came to be.

Nor is this a history of the globalisation of fossil capital in general. It simply embeds the rise of fossil capital and associated processes in Egypt within its global context, this being one of the earliest cases of fossil capital penetrating regions and countries outside its birthplace, England, during the Industrial Revolution. It is also not a history of the means or modes of production. I approach the social history of production technology from the point of view of two categories, energy and transitions, and I attempt to study how the former evolved dialectically with and within the latter.

Energy regimes applied in agrarian and industrial production across the decades (1820s–70s) in Egypt evolved in a non-linear manner, with steam power taking two steps forward, then one step back.[2] It burst into factories, then moved away to fields in the form of steam ploughs and pumps, before evolving again as production lines inside factories built anew. This peculiarity of energy, or the question of energy transition, makes sense only once we locate it within the fabric of broader and multiple transitions. These transitions were non-linear, gradual (although occasionally abrupt), and grounded in complex processes on many scales, of which we can name some. A transition from in-kind taxation of the producers to monetised taxation; from empire-building, by Mehmed Ali, geared towards expanding monopolies on mass-produced commodities for trade, to growing dependency (1850s–60s) riding on loans taken from European capital to grow and process mono-crops, of first cotton and then sugar; from indentured corvée labour, to a moment when 'free' wage labour of the same peasants started to take shape (mid-1870s); from universal tax-farming as the hegemonic extractive mechanism in rural Egypt, to the origination of concepts that

2 While 'fossil capital' (see: Andreas Malm, *Fossil Capital: The Rise of Steam Power and the Roots of Global Warming*, London: Verso, 2016) is an energy lens I adopt for researching and writing this history, I acknowledge the broad scholarship on questions related to energy from Arab-region historians and how useful it was while designing this project. This includes work on labour and animal power in the early modern Ottoman Empire (Alan Mikhail, *Under Osman's Tree: The Ottoman Empire, Egypt, and Environmental History*, Chicago and London: University of Chicago Press, 2017) and earlier (Richard Bulliet, *The Camel and the Wheel*, New York: Columbia University Press, 1990); labour history of miners and coalheavers by Donald Quarteart and John Chalcraft; oil in state formation by Toby Craigh Jones; land and oil concessions and oil production as techno-social processes by Katayoun Shafiee; social history of oil workers and labour movements by Timothy Mitchell, Touraj Atabaki, Maral Jefroudi, Peyman Jafari, and Kaveh Ehsani; the 'oil town' and urbanism in petrostates, by Nelida Fuccaro, Claudia Ghrawi, Rasmus Elling, Mona Damluji, Arbella Bet-Shlimon, Reem Alissa, and Farah Al-Nakib.

included 'industrialisation' and 'thermodynamics', with the rise of the production line in the late 1870s. I will return to and detail further this panoramic view in the last section of the Introduction.

The manner in which the question of energy regimes fits into this broader fabric of transitions is the prime topic of this book. Energy is central, but it is not the only or first priority. Energy is a science that originated in the nineteenth century,[3] and became an epistemic point of departure for many scholars who invite us to look at energy itself, in neat economic models, disconnected from historical change.[4] However, it is not a transhistorical category, as conceptual histories of energy show.[5] It needs to be thoroughly historicised as embedded in all sorts of social, political, and technological changes, as Fred Albritton Jonsson argues.[6] Before outlining the ways in which energy moved back and forth – as this book demonstrates – I first need to explain how I define and deploy

3 Cara Daggett, *The Birth of Energy*, Durham, NC: Duke University Press, 2019; Mathew Shutzer, 'Energy in South Asian History', *History Compass* 18:12, 2020.

4 Energy transitions – in this strand of the literature – are depicted exclusively as bounded technical and market-related processes, stories of innovation, and the move from niche to general markets, even while acknowledging that it is usually a 'multi-actor' process. Although useful for designing energy policies, these insights seem to hardly appreciate the complicated histories 'energy' is usually enmeshed in. See Karoline Rogge et al., 'Conceptual and Empirical Advances in Analysing Policy Mixes for Energy Transitions', *Energy Research and Social Science* 33, 2017, 1–10; Roger Fouquet and Peter J. G. Pearson, 'Past and Prospective Energy Transitions: Insights from History', *Energy Policy* 50, 2012, 1–7; Robert C. Allen, 'Energy Transitions in History: The Shift to Coal', *RCC Perspectives* 2, 2013, 11–16.

5 For example, Ivan Illich, 'The Social Construction of Energy', in *Beyond Economics and Ecology: The Radical Thought of Ivan Illich*, edited by S. Samuel, London: Marion Boyars, 2013, 105–23; Phillip Mirowski, *More Heat Than Light: Economics as Social Physics, Physics as Nature's Economics*, New York: Cambridge University Press, 1989; Anson Rabinbach, *The Human Motor: Energy, Fatigue and the Origins of Modernity*, New York: Basic Books, 1990; Daggett, *The Birth*.

6 He addresses the 'neglect of energy' in historical scholarship by asking helpful questions: how different are fossil fuels from other commodities? 'What social, technological, and political forces paved the way for the transition to coal, oil, and gas?' (see: Fredrik Albritton Jonsson, 'Anthropocene Blues: Abundance, Energy, Limits', in *The Imagination of Limits: Exploring Scarcity and Abundance*, edited by Frederike Felcht and Katie Ritson, *RCC Perspectives* 2, 2015, 55–63). But, naturally, we can also ask how these forces changed under the impact of the transition to fossil fuels. On Barak, through different approaches that adopt a 'counter-energy' view, does this successfully in the modern history of the Arab region (see: On Barak, *Powering Empire: How Coal Made the Middle East and Sparked Global Carbonization*, Berkeley: University of California Press, 2020).

certain concepts in regards to the social, political, and technological forces that I will focus on during these transitions. These concepts are capitalism, industrial capitalism, fossil capital, and the Capitalocene.

I use the term 'capitalism' based on a concise definition: the generalised production of commodities for their exchange value exclusively, where commodity production is a process geared solely to create surplus value for exchange. This is the rule of capital, or capitalism, where capital survives by the continuous cycles of creating and realising surplus value. Capitalism is a social relation between persons mediated through things – a capitalist and labourers and, between them, property relations.[7]

Industrial capitalism refers here to the period beginning when the generalised production of commodities for their exchange value started to take place through fossil-fuelled factories with production lines. In this book, I refer to the integration of Egypt into the industrial capitalism of Europe as a provider of material inputs and processed agrarian commodities across the nineteenth century.

Fossil capital is a relation and a process. It is, first, a triangular relation between capital, labour, and fossil fuels, by which capital's exploitation of labour occurs through the consumption of fossil fuels, and where surplus value accumulates through the burning of fossil fuels into carbon dioxide. It is, second, a process through which value accumulates by burning more and more quantities of fossil fuels, that is, the more that capitalism (the rule of capital) extends in time and space, the more fossil fuels are burned.[8] We will see, across this book, how European fossil capital found ways to extend to Egypt and integrate it into global – industrial – capitalism through successive moments. To use this term is to remain sensitive to the political ecological nuances of the history of how fossil-fuelled technologies evolved in nineteenth-century Egypt, being a history, ultimately, of the rise of global warming from a particular place and time.

Now, after briefly presenting how I use these concepts that are relevant to the questions of transition that this book tackles, it will be useful to keep in mind an overarching concept, an aggregate that captures

7 Both defining characteristics of capitalism, from Michael R. Krätke's reading of *Capital, Vol. III*. See Michael R. Krätke, 'Capitalism', in *The Marx Revival: Key Concepts and New Interpretations*, edited by Marcello Musto, Cambridge: Cambridge University Press, 2020, 6–8.

8 Malm, *Fossil*, 290.

many of the multiple processes that formed the techno-energetic transition. This is what can be described as the epochal global rise of the Capitalocene, being the umbrella concept that holds sway over the processes and concepts listed here, as well as others.

The Capitalocene is a historical-materialist conception of the Anthropocene.[9] Through it, climate historians and other scholars of the social sources of global warming may describe, analyse, and make sense of how we ended up with our heated planetary condition, and formulate perspectives on what comes next. 'Capitalo-' (capitalist metabolism of society) thus creates a 'cene', a geological epoch (in nature), instead of blaming humans at large (Anthropocene) for the historical and future processes and consequences of global warming. It is through this political and ecological concept that this book seeks to prevent the history covered from being rendered as the usual sorry saga that fits all too well with the current climate dithering. It seeks to cast a net broad enough to capture the moment when the current epoch was born.[10]

How does this book view the history of the origin of the Capitalocene from the standpoint of precolonial nineteenth-century Egypt? First, by definition (given the Capitalocene), it is a socioecological history of steam power and technology rising through pre-existing economic and social formations across varying scales – local, regional, and global – and actively shaping and becoming shaped by these broader transitions. Second, it is part of the global history of empire, primarily north-western

9 David Ruccio, Donna Haraway, Jason W. Moore, Andreas Malm, and others developed the concept with different nuances, the common denominator being a critique of the overly generalising and responsibility-diffusing Anthropocene. And there are other terms (for example, Anthrobscene, econocene, manthropocene) that aim to break with the 'all of humanity is responsible' narrative. In this book, I use the term departing from how Andreas Malm used it in *Fossil Capital*, simply 'the geology not of mankind, but of capital accumulation'. See Malm, *Fossil*, 391. For the lineage of the concept, see Jason W. Moore, *Anthropocene or Capitalocene? Nature, History, and the Crisis of Capitalism*, Oakland: PM Press, 2016, 5–6. Christopher Bonneuil and Jean-Baptiste Fressoz provided a lucid account of the Capitalocene's intellectual lineages; see Christopher Bonneuil and Jean-Baptiste Fressoz, *The Shock of the Anthropocene: The Earth, History, and Us*, translated by David Fernbach, London: Verso, 2016, Chapter 10.

10 Following Andreas Malm and Alf Hornborg, 'The Geology of Mankind? A Critique of the Anthropocene Narrative', *The Anthropocene Review* 1, 2014, 62–9.

Europe during the Industrial Revolution, at the moment when high imperialism (1870s–1940s) was taking shape.[11]

In other words, this book presents a socioecological history of energy and technology in production in Egypt before the British occupation of 1882. One purpose is to contribute to ongoing studies of how the main machinery that unleashed global warming found its way into regions outside Europe.

Historicising the Capitalocene, Energy, and Technology: To Study All History Anew

What can historians do to push the limits of our understanding of climate change, which by definition is a historical phenomenon? Christophe Bonneuil and Jean-Baptiste Fressoz suggest one agenda: the Anthropocene, they say, 'this forceful, even violent, return of the history of the Earth into world history creates a new human condition and requires us to reintegrate nature and the Earth system at the heart of our understanding of history'.[12] It follows that this reintegration of nature in history should be, they continue, a more explanatory history of the Anthropocene – as opposed to one that studies the history of climate as a chronicle of changes in the biogeochemical cycles – zooming in on 'actors, institutions and decisions' that were the prime movers of the climate crisis.[13]

This is a convincing agenda that needs some further unpacking. First, the historical inquiry into global warming should be about actors, institutions, and decisions, about the social realm. An inspiring framing here is one that the environmental sociologist Kari Marie Norgaard suggested for her field, as she followed the concept of 'the Sociological Imagination'.[14]

11 For the concept of high imperialism, see Corey Ross, *Ecology and Power in the Age of Empire: Europe and the Transformation of the Tropical World*, Oxford: Oxford University Press, 2019, 7.

12 Bonneuil and Fressoz, *The Shock*, Chapter 2.

13 Ibid. Andreas Malm separately crafted a similar agenda – history in climate instead of climate in history – under the question, 'Who lit this fire (of global warming)?' See Andreas Malm, 'Who Lit the Fire? Approaching the History of the Fossil Economy', *Critical Historical Studies* 3:2, 2016.

14 C. Wright Mills, *The Sociological Imagination*, Oxford: Oxford University Press, 1959, 3–7.

A sociological inquiry into nature/society phenomena, she suggests – based on this approach – can combine levels of analysis ranging from the macro (structural, as in the political-economic realm) via meso (for example, the cultural and psychological) to micro (for example, individual actors).[15] Another disciplinary point of departure compatible with such an approach is historical political ecology, as outlined by, for instance, Diana K. Davis: to gear the historical analysis of global warming towards understanding the actors and processes that started it.[16] Methodologically, in terms of the analytical framing of 'technology' – which this book follows closely as a cause of global warming in combination with actors, institutions, and decisions – the scholarship of Alf Hornborg and Timothy Mitchell offers, if seen together, a useful analytical space. The former seeks to socialise the history of technology, to see a social and human history embodied in the technological.[17] The latter is occupied with the inverse: to end the exclusion of the technical and natural (not technology per se) from analysing the social.[18] Between these lenses, a dialectical space opens up, a stage on which we can account for the technical in the social, and the social in the technical, simultaneously, across causal chains. The point is that this history should be a social history, placing energetic and technological change in the contexts of the social relations through which it unfolded.

Second, since it is a social history of certain episodes of human metabolism with the rest of nature through energies and technologies of production, historical materialism and the focus on capitalism shall be central. This is because capitalism – the generalised production of commodities for their exchange value through subsuming labour and nature – is the overarching structural relation shaping human

15 Kari Marie Norgaard, 'The Sociological Imagination in a Time of Climate Change', *Global and Planetary Change* 163, 2018, 171–6. See also: Kari Marie Norgaard, *Living in Denial: Climate Change, Emotions, and Everyday Life*, Cambridge, MA: MIT Press, 2011, 1–12.

16 Diana K. Davis, 'Historical Political Ecology: On the Importance of Looking Back to Move Forward', *Geoforum* 40:3, 2009.

17 Alf Hornborg, *Global Magic: Technologies of Appropriation from Ancient Rome to Wall Street*, New York: Palgrave Macmillan, 2016. See 'Introduction'.

18 Timothy Mitchell, *Rule of Experts: Egypt, Techno-politics, Modernity*, Berkeley: University of California Press, 2002, Chapter 1.

metabolism over the last few centuries.[19] A socioecological history of global warming must be one of capitalism and how it extended globally. Following this central point, which I develop further in the next section, the concept of the Capitalocene offers a promise of acting as a conceptual umbrella covering the many transitions of nineteenth-century Egypt, as well as the developing social relations of labour and property, and the tendencies of capitalist expansion and the reproduction of cycles of accumulation across global, regional, and local scales.

This does not entail, however, that capitalism or the Capitalocene shall remain, for the purposes of this book, static concepts imposed from above. This is the third point: what happens when historical-materialist scholarship critically engages the concept of the Capitalocene as an historical analytical category? While the Anthropocene offers a most generic temporal range for how nature returned forcefully to inhabit human history, the history of capitalism – in nature – does not seem to be similarly complete in its universality. Michael Krätke tells us that Marx's view of the historical development of capitalism, by the 1880s, was that it had never been a unilineal process, hence the unfinished nature of Marx's *Capital*.[20] Marx acknowledged, in the 1870s, that western European capitalism was just 'an ideal average' for capitalism.[21] This is more a problem than a solution for studying the non-linear development of capitalism. Hence,

> Engels, who claimed to have no maxim, formulated [a solution] that he and Marx followed as soon as they designed the research programme for the '*materialist conception of history*': 'The whole of

19 On the conceptual history of capitalism in combination with metabolism, see John Bellamy Foster, *Marx's Ecology: Materialism and Nature*, New York: Monthly Review, 2000; metabolic rift and capitalism: John Bellamy Foster, Brett Clark, and Richard York, *The Ecological Rift: Capitalism's War on Earth*, New York: Monthly Review, 2011; metabolism and environmental history: Steven Stoll, 'A Metabolism of Society: Capitalism for Environmental Historians', in *The Oxford Handbook of Environmental History*, edited by Andrew C. Isenberg, Oxford: Oxford University Press, 2014.

20 On Marx's thoughts and studies about the Global South and capitalism outside Europe during the later decades of his life, see Kevin B. Anderson, *Marx at the Margins: On Nationalism, Ethnicity, and Non-Western Societies*, Chicago: Chicago University Press, 2016; Marcello Musto, *The Last Years of Karl Marx: An Intellectual Biography*, Stanford: Stanford University Press, 2020.

21 Michael R. Krätke, 'Marx and World History', *International Review of Social History* 63, 2018, 91–125, endnotes 82, 83.

> history must be studied anew'.[22] This is also true for the long and complicated history of capitalism that Marx's general theory has to prove itself against.[23]

The whole of history must be studied anew. In other words, if one is after the history of capitalism as central to the history of global warming, then the very concepts of capitalism and the Capitalocene have to be grounded in the respective histories of their distinct moments. It is in this sense that this book historicises the Capitalocene, energy, and technology from nineteenth-century Egypt. Nothing is taken for granted. It shall be a highly empirical history that will be – as far as possible – one step ahead of theory.[24]

The core argument here is that the history of the nineteenth century is the history of the beginnings of global warming. Global warming is a planetary phenomenon, an existential crisis with looming and already realised breakdowns of all sorts. However, it came about as a result of developments that began in the nineteenth century, first and foremost the large-scale combustion of fossil fuels, diffused across the world in the form of steam power. It follows that the nineteenth century must be reconceived and studied anew as a history of global warming. It is then up to the historians of the Anthropocene – better, the Capitalocene – to design and delimit their questions within this overarching problematic.

Historical Materialism and Global Warming

The whole of history must be studied anew, and it follows that the thorough revisiting of this history, geared to accounting for the rise of global warming, must be undertaken. This is in order to see and explain moments, articulations, and processes of human metabolism in nature, which together form what we can, in hindsight, plot as a history of the Capitalocene. But what is the Capitalocene in analytical terms? How is

22 Engels to Conrad Schmidt, 5 August 1890 (MEW 37, p. 436; MECW 49, p. 8), cited in Krätke, 'Marx'.

23 Krätke, 'Marx'.

24 Jairus Banaji's call for putting history before theory is an inspiration for this methodological point of departure. Jairus Banaji, *Theory as History: Essays on Modes of Production and Exploitation*, Leiden: Brill, 2010.

it related to the history of agrarian production technology in nineteenth-century Egypt? In this book, I deal with the questions of nature and capitalism (the ingredients of the Capitalocene) by following a logic of gradation.

Gradation, or 'matters of degree' in thinking about the ontological question of nature and society, helps avoid the pitfalls of two major sets of epistemological positions that have emerged in response to the challenges of the Anthropocene. On the one hand, there are constructivist approaches that see nature as a construct of human society. Constructivist normative attacks on the foundations of modernity, including Western rationality and culture, seek to dissolve binaries, as in the important work of Val Plumwood.[25] Discursive postmodernist constructivism investigates human representations of nature, but does not ask what nature is.[26] Strictly materialist constructivism, meanwhile – as exemplified by Steven Vogel – argues that nature has been literally constructed by human actions and discourses since the rise of *Homo sapiens*.[27] By collapsing nature into society, however, these approaches leave us with analytical tools that are too blunt to shed light on specific mechanisms of appropriation and social injustices. On the other hand are positions (such a hybridism, post-humanism, and new materialism) that seek to avoid anthropocentrism by invoking the agency of things, and which deliver catchy headlines and keywords that frame the events of history through powerful, haunting imagery – yet without venturing towards questions of how and why. An example of such scholarship on the Arab region, which does factor in nature and materiality, is On Barak's *Powering Empire*, which leaves us with 'Coalonialism' as its framing of the rise of steam power and empire.[28] This conclusion, that coal burning 'intensified' exploitation, leaves us back where Marx began in his analysis of machines.[29] In sum, constructivist positions collapse nature into society, while new materialist or more-than-human positions collapse

25 Val Plumwood, *Environmental Culture: The Ecological Crisis of Reason*, London and New York: Routledge, 2002.

26 Noel Castree, *Making Sense of Nature*, London and New York: Routledge, 2013.

27 Steven Vogel, *Thinking Like a Mall: Environmental Philosophy After the End of Nature*, Cambridge, MA: MIT Press, 2016.

28 Barak, *Powering*.

29 Karl Marx, *Capital: A Critique of Political Economy, Vol. I*, translated by Ben Fowkes, London: Penguin Books, [1867] 1976, Chapter 15.

society into nature. Neither leaves us with analytical tools that are sharp enough to reconstruct a synthesis of the material combined histories of society and nature.

The fascination with shattering modernist analytical categories and binaries of all sorts across the various fields of social science since the 1990s pushes for cosmologies and life-worlds other than the one built by modernity, but the destructive life-world of capitalism needs to be understood on its own terms if humans are to continue to exist on this planet. This claim is inspired by Philippe Descola's *Beyond Nature and Culture*, in which he suggests how cosmologies are formed. The Achuar in the Amazon rainforest, he shows, built a cosmology where non-human beings are connected to them by marriage and blood relations. Agrarian products are blood relatives to women, and hunted game are the in-laws of male hunters. Everything in the Achuars' biome, the human and non-human, falls within a hierarchy in which humans are at the top and non-human beings occupy lesser levels: animals, plants, and spirits of the forest, some closer to humans than others. Humans communicate together through language; humans and non-humans communicate in trances, hallucinations, and dreams. Like the new materialist and post-humanist approaches, the ontological grounding of the Achuar knows no clear borders between the human and non-human. In their cosmology, the world is inhabited by extremely diverse beings with all sorts of ways to connect and communicate.[30] The Cree of northern Canada, who inhabit a less diverse world, view the fewer mammal species they live with as respected 'nations' of sorts, and as such hunting game is rationalised.[31] In both cases, the agency of the non-human, the material at large, is in dictating a worldview that makes sense of humans' metabolism in nature.

Unlike the Achuar and the Cree, we know that the generalised production of the commodity devastates nature through the inherently destructive byproduct of global warming. Such is the simple and clear base of the modernist metabolism of humans in nature. Historicising the origination of this specific and bounded metabolism is a responsibility of historians, and, while more-than-human approaches can offer

30 Philippe Descola, *Beyond Nature and Culture*, translated by Janet Lloyd, Chicago: University of Chicago Press, 2013, Chapter 1.

31 Ibid.

powerful and creative analytical units that destabilise anthropocentrism, capitalism, with its human actors and their decisions, needs to be kept in focus. A powerful critique of capitalist metabolism, historical materialism, can offer a grounded analytical framing capable of explaining human-nature metabolism towards devastation (global warming) by fleshing out the history of the globalisation of fossil capital (and capitalism as embedded in nature) across time and space. This is what I mean by matters of degree, towards seeing and analysing the human-nature dialectic across time, and the rise of capitalism and fossil capital in pre-capitalist formations, as two separate but linked landscapes.

Nothing is new in this perspective I offer. The analytic of landscape is borrowed from Simon Hailwood's *Alienation and Nature in Environmental Philosophy*, in which he suggests, in a pluralist approach, that a landscape – a village, a town, a painting, a text – belongs to both nature and society, following a logic of gradation. A village is non-human insofar as non-human beings and things exist in it, and it is human insofar as humans inhabit and change it. A painting is non-human insofar as it is formed of chemical colours and canvas or paper, and human insofar as it is the product of the painter's agency, and so on and so forth.

Andreas Malm, based on the concept of landscape, developed the analytic of nature/society combinations. He starts from the view that social and natural relations are inevitable: and nature and society are continuous in the material world.[32] However, humans are unique in that they can take decisions about how to organise their metabolic exchange with nature, which – under capitalism and the logic of continuous accumulation – can have planetary-scale consequences. This metabolic peculiarity of humans in nature is expressed in the concept of nature/society combinations, and it can be seen in any sort of human activity that has to bend to social/human laws of motion as much as it does to biogeophysical rules. This is the ontological position from which my questions, analysis, and thinking about historical causalities are presented in the succeeding chapters. The agency of nature, the

32 Malm also drew on Kate Soper's realist definition of nature: Kate Soper, *What Is Nature? Culture, Politics and the Non-human*, Oxford and Cambridge, MA: Blackwell Publishers, 1995. Andreas Malm, *The Progress of This Storm: Nature and Society in a Warming World*, London: Verso, 2020, 60.

non-human, as I will show shortly, manifests itself in defying the subsumption of nature to the rule of capital, but this defiance flows from human causes. As Chapter 2 demonstrates, for example, decisions taken to redistribute land in Egypt in the late 1830s – often defined in the literature as the moment that a landed class was created – led to the 'natural' disaster of the cattle plague, which, in its turn, made it more possible to turn in later years to steam-powered irrigation: the dialectical dance of nature/society combinations (Malm's terms) in the nature/society landscape (Hailwood's).

The same logic of 'matters of degree', the landscape, can be gleaned from Peter Hill's rethinking of the pre-capitalism/capitalism transition in terms of 'nodal thinking': globally expanding capitalism found its ways to specific pre-capitalist social formations across time, akin to a network turning on more and more nodes in different proportions. And the model is fractal; inside a node on the global network, local and transregional nodes of capitalism start to be turned on: railways built, boom towns turning to cotton, a proliferation of credit and borrowing on many levels, and the creation of factories and modern production lines summoning a working class.[33] Following Hill, Egypt in the 1820s–70s then itself becomes a node, a landscape across which further capitalist nodes are turned on. It is pre-capitalist insofar as labour is massively appropriated coercively (not yet appropriated via Marx's real subsumption of labour), and capitalist insofar as European loans (financial capital) are taken out to import steam irrigation infrastructure, to which coercive pre-capitalist labour was tied. It was also becoming a node of global capitalism insofar as industrial production lines started to be assembled and their steam engines ignited, and insofar as European fossil capital started to penetrate the Delta and Nile valley, bringing colonialism on its heels.

This landscape in transition from pre-capitalism to capitalism can be seen as a stage on which techno-energetic transitions are historicised, allowing for a range of focus and avoiding getting entangled in the problem of constantly defining what is capitalist and what is pre-capitalist. The nature/society combinations landscape is useful in this book for laying out a socioecological history of energy and technology in

33 Peter Hill, *Utopia and Civilisation in the Arab Nahda*, Cambridge: Cambridge University Press, 2019, 20–3.

agrarian production. The reason for this choice is that Marxist scholarship that seeks to locate capitalism in Egypt's modern history tends to be *longue durée*, abstract (in the sense of lacking new empirical work while arguing for novel analytical insights), theory-oriented, and less focused on concrete historical cases of change in the means of production.[34] Studying history anew, by offering a synthesis of agrarian production energies and technologies in their connections to European fossil and financial capital, as this book does, can help us see ignored junctures and moments of change, relevant to Egypt's modern history as much as to the history of capitalism's destructive career in nature.

Egypt Ignited, in theoretical terms, is therefore an attempt to start uncovering a long process of the rise of fossil capital in connection to the rule of capital (capitalism) in Egypt from the viewpoint of the techno-energetic setup of the means of production, and an attempt to explore the 'global' in global warming, by starting to account for the rise of the Capitalocene from a particular place in the Global South. Simultaneously, the metaphor of ignition is about how this history of technology and energy intersects with questions about the ways in which capitalism was articulated with the Egyptian pre-capitalist social formation. This is how the analysis of fossil capital as deployed here can be useful in the quest to understand the moments and nodes of capital in Egypt's history. Concretely, I claim that showing how Egypt was ignited – a metaphorical steam engine coming to life across fifty years – is crucial to understanding the place of capitalism in Egypt's modern history across the nature/society landscape it created.

34 A recent overview on capitalism in modern Egypt: Aaron Jakes and Ahmed Shokr, 'Capitalism in Egypt, Not Egyptian Capitalism', in *A Critical Political Economy of the Middle East and North Africa*, edited by Joel Beinin, Bassam Haddad, and Sherene Seikaly, Stanford: Stanford University Press, 2021. Another useful *longue durée* and more analytical but still abstract overview (with a focus on labour subsumption) is: Jelle Versieren and Brecht de Smet, 'Lost in Transitions? Feudalism, Colonialism, and Egypt's Blocked Road to Capitalism (1800–1920)', *Historical Sociology* 35:2, 2022, 200–21. Earlier important work on labour history in Egypt was blind to early instances of labour subsumption, as it searched for the origins of a capitalist 'working class'. Thus, Ellis Goldberg dismissed the sugar factories of the 1870s as 'too small' (they were massive), and Joel Beinin thought that the only mechanised industry in the 1860s–70s was cotton ginning. So in their work and that of others, modern labour history tended to start in the late 1880s–90s. See Ellis Goldberg, *Tinker, Tailor and Textile Worker: Class and Politics in Egypt, 1930–1952*, Berkeley: University of California Press, 1986; Joel Beinin, *Workers and Peasants in the Modern Middle East*, Cambridge: Cambridge University Press, 1987.

Outline of the Book

With this agenda, the main thrust of the book is a detailed revisiting and analysis of nineteenth-century Egypt with a focus on energy and technology, across two stages. First, the rising then falling curve of the predatory but frustrated empire-building project of Mehmed Ali that sought to compete with Britain, thwarted through the massive imbalance in coal endowments between the two. Second, originating from this moment, the long downwards curve from 1850 into the 1870s, Egypt's gradual descent into a dependency on fossil-powered western Europe and its simultaneous integration into global capitalism. With this agenda, the book reconsiders Egypt's precolonial modern history. This is the source of the definition of the Capitalocene as it was extended to precolonial nineteenth-century Egypt: the Capitalocene, I argue, came to Egypt in the form of successive cycles of crises, bringing fossil-fuelled technologies of production to the Nile, purchased with loans from European financial capital. This process entrenched western European fossil and financial capitalism on the Nile, followed by outright colonialism through the British military occupation of 1882.

The book offers a synthesis of the precolonial nineteenth century in Egypt: the rise and fall of Mehmed Ali's empire dream, followed by decades of growing dependency on financial European capital and fossil-fuelled technology into the 1870s. The final short section offers a brief intervention on cultural and conceptual aspects of this history, spanning these long decades and spilling into the early twentieth century. The aim of this last part is to begin to account for how capitalism was eventually translated and adopted as an ideological project for the Egyptian ruling class, and how and why the direct producers might have reacted to steam technology and its capitalist logic through rationalising it as magic.

In Chapter 1, I set the scene for the industrial production of mass commodities during the earlier decades of Mehmed Ali (1820s–30s), conditioned by the scarcity of coal. Chapter 2 charts the crucial moment of the 1840s, when the Pasha's imperial project, frustrated by the sustained absence of cheap locally acquired coal, gave way to long decades of dependency on expanding Europe-centred capitalism. The chapter follows the Pasha's conquests of Syria across the 1830s and argues for putting the search for coal at the centre of analysis. It relates the cattle epizootic of the early 1840s to the freeing of trade enforced by the British, the formation of a landed class in Egypt, and the abrupt collapse of the Pasha's industrialisation over the decade.

Chapter 3 follows how fossil-fuelled technology started to trickle into Egyptian agrarian production only after the fall of Mehmed Ali's industrialisation. When steam technology began to be imported, it came to the large farms of the rich landed class formed during the years of Mehmed Ali's power crisis (early 1840s). Alongside the land, it provided the agrarian raw materials necessary for European industrial centres. The new energy and technology on these farms were needed to fix the deterioration of the irrigation infrastructure, and to produce white sugar, which, in order to meet rising consumer expectations at the time, required steam power.

Chapter 4 goes into a decisive moment for Egypt's growing dependency on Europe and integration into global capitalism. It revisits the age of Sa'ed Pasha (1854–63), who monetised state taxation and the agrarian economy. It focuses on the spread of steam technology in cotton ginning, and how this technology spurred more penetration of credit fractally across the various levels of Egypt's agrarian economy, from the small farmer to the Viceroy and the rich landed class.

I start Part II, which takes us through the reign of Ismail Pasha, with a theoretical intermission (Chapter 5) that challenges the mainstream narrative of Khedive Ismail as a wasteful oriental prince who was deceived by premeditated European designs for colonising Egypt by entangling it in irrational borrowing. Here, I complement the nature/society combinations analytic deployed in Part I with an analytical framing that intervenes in what seems to have been the most epochal moment of transition in Egypt's uptake of fossil-fuelled technology and the penetration of global capitalism.

Chapters 6 and 7 present and analyse how steam technology came to Egypt in a massive and unprecedented wave in the wake of a crisis in 1863 – the cotton famine/boom, associated with a second cattle epizootic on the Nile. The Khedive, who had consolidated his landholdings after he ascended the throne, soon bought the locomobile, or the steam cultivator, a 6–20-horsepower steam engine used for ploughing and irrigation, mainly from England. He also consolidated his power by dictating who possessed steam power in the country and for what purposes. Unlike the epizootic of the early 1840s, which led to the collapse of industrialisation, the 1863 epizootic seems to have spurred the uptake of steam power in agricultural production – irrigation, cotton ginning – which further entrenched European fossil and financial capital, through machine exports and loans. By the mid-1870s, more lands were irrigated by steam than by animal power, while Egypt was on the verge of declaring bankruptcy.

Chapter 8 shows how the sudden mass adoption of the new technology in the mid-1860s unleashed unforeseen socioecological contradictions, focusing on the local uprising of Qau in 1865. Chapter 9 follows how this explosion of steam-powered technology in agrarian production soon consolidated knowledge networks between London, Odessa, and Cairo, through which biomass of all sorts, found on the field, became the alternative fuel for the steam engines and boilers, given the chronic scarcity of coal on the Nile. One would not expect such a twist in the narrative of a history tracing fossil-fuelled technology, but this shows how the scarcity of coal might have inspired more expansion in modern, fully mechanised agrarian production, leading to even more coal imports and more burning of fossil fuels in the long run. Overall, Part II offers an account of how fossil-fuelled technology spread on the Nile through the 1860s and 1870s, including the infrastructures – and alternative fuels – for this phase of the extension of capitalism to agrarian production in Egypt.

Putting the question driving this book to learned Egyptians from the period and the masses of peasants, Chapter 10 takes us into the realms of ideology and culture. The most prominent Egyptian intellectual in the mid-nineteenth century was Rifa'a al-Tahtawi. Here, his thoughts on macroeconomics and political economy are examined to explore his conception of capitalism. In it, what he coined *al-industriya* – from the French word *industrie* – motivates the generalised production of commodities, geared towards international trade. Another term for the latter, he thought, would be 'industrial colonialism'. Peasants, on the other hand, had a different appreciation of the steam engine and its absurd industrial accidents. To them, it represented a form of black magic. It was a technology they received with uneasiness and suspicion once it showed its dark, destructive face – a perception of eerie relevance to present concerns regarding global warming. For peasants in late nineteenth-century and early twentieth-century Egypt, something seemed inherently wrong with fossil-fuelled technologies and the new epoch they heralded. With this I conclude the book.

PART I
The Frustrated Empire

Addressing me, he said, 'Welcome to Egypt.' I bowed. 'I have sent for you to travel in my country to search for minerals, but particularly for coal, of which I stand much in need. [. . .] I have a manufactory at Boulak, which I wish you to see; and I want to enlarge it, when you have found coal for me, [so] that I may make guns and steam-engines, and feel myself independent of England.'

Mehmed Ali Pasha, speaking to John Petherick,
a British mining engineer, in 1845
(John Petherick, *Egypt, the Soudan and Central Africa*)

1

The Industrial Revolution That Never Happened (1820–39)

In terms of motive power, the history of Mehmed Ali's push for industrialisation is centred on a contradiction. His factories suffered from a scarcity of steam power and water power, but despite this basic fact casting doubts on industrialisation in the modern sense,[1] this was Egypt's first and most spectacular example of industrialisation across the entirety of the nineteenth century. Mehmed Ali is often referred to as 'the founder of modern Egypt',[2] and the modernity being referred to is usually associated with industrialisation, among other defining features.

However, underestimating the Pasha's industrialisation on account of lacking in modern prime movers[3] – not without reason, obviously

1 Roger Owen, for example, found industrialisation proper at odds with the outlook in Britain at the time: a handful of steam engines in 1820s–30s Egypt versus at least 10,000 steam engines in England in 1822. See Roger Owen, *The Middle East in the World Economy*, London: I. B. Tauris, 2002, 72.

2 Timothy Mitchell traced the use of this phrase for Mehmed Ali to the 1930s, from Henry Dodwell, a British historian of colonial India (see Henry Dodwell, *The Founder of Modern Egypt: A Study of Muhammad Ali*, Cambridge: Cambridge University Press, 1931). It was already being used, however, as early as 1925 in the Egyptian press (*Al-Musawar*, 30 January 1925) and as such seems to have a longer history. This notion (along with Dodwell's book) has been challenged by historians since at least the 1970s. But the idea of Mehmed Ali as the 'founder of modern Egypt' lives on in travel websites and brochures, and popular biographies.

3 For example, Khaled Fahmy criticised earlier framing of Ali's factories as 'a revolution in industry' on account of having too few steam engines, the absence of trained labour, and 'inherent bureaucratic complications'. In the first two factors he follows

– also runs against evidence on the grand output of the factories. For example, engineer Alexander Galloway, a major London employer in the steam engines business in the 1820s and 1830s, told the House of Commons in 1837 that the Boulaq foundry, established in the 1820s by his son, was 'one of the first foundries of the world'.[4] It hired around 5,000 workers and dozens of Egyptian engineers, beside some forty to fifty British experts.[5] Along with other foundries and arsenals, towards the late 1830s, this industry produced over twenty naval vessels, including nine warships with more than a hundred cannons each. By the mid-1830s, just a single spinning mill, from among more than thirty across Egypt, met all local demand for cotton thread with a margin for export.[6]

This chapter does not evaluate this industrialisation. Instead, it shows, from the viewpoint of energy, a unique configuration of large-scale production technology, which can be best understood as an amalgam of 'modern' machines disconnected from steam engines, and powered instead by oxen and men.[7] The argument I make is that this hybrid and unique configuration of industrial production, of humans, animals, and modern steam-deprived machines, was set up purposefully as a provisional measure until carboniferous sources for igniting steam engines could be found. In other words, Mehmed Ali must have been fully aware of the fossil basis of steam engines and took a decision to postpone the

---

Roger Owen's subtly presented doubts at the whole industrial enterprise of Mehmed Ali. See Khaled Fahmy, *All the Pasha's Men: Mehmed Ali, His Army and the Making of Modern Egypt*, Cairo: AUC Press, 2002, 20; Owen, *The Middle East*, 72.

4 House of Commons, *Report from the Select Committee on Steam Communication with India*, London: House of Commons, 1837, 66.

5 'Parliamentary Report on Steam Communication with India', *The Mechanic's Magazine, Museum, Register, Journal, and Gazette* 28, 1837, 186.

6 Owen, *The Middle East*, 70–1.

7 A parallel to this 'hybrid' in motive power of the factories is the hybrid organisation of labour at the Pasha's factories, which, arguably, leaned heavily towards the old guild system. Perhaps the most detailed and thorough body of work on the guild system at the factories of the early to mid-nineteenth century is Pascale Ghazaleh's, which relied on rarely accessed primary material from the *fabriqat* (manufactories) records of Mehmed Ali, around the year 1823. See Pascale Ghazaleh, 'Masters of the Trade: Crafts and Craftspeople in Cairo, 1750–1850', *Cairo Papers in Social Science* 22:3, 1999; Pascale Ghazaleh, 'Manufacturing Myths: Al-Khurunfish, a Case Study', in *Money, Land and Trade: An Economic History of the Muslim Mediterranean*, edited by Nelly Hanna, London: I. B. Tauris, 2002, 124–35.

systematic importation, or manufacture, of steam engines for his industries, until he could acquire cheap local stocks of coal.

The chapter begins with a selective survey of the technical setup of the Pasha's factories in the 1830s, and their prime movers. This is to show the lay of the industrial landscape at its strongest point in the first half of the century, and to make sense of how the few steam engines seen at the time in the factories were deployed and integrated. Next, I move back to the 1820s, when Mehmed Ali first entertained the idea of the steam engine and its fossil fodder and proceeded to import some such engines. Finally, I examine other prime movers the Pasha considered: wind and water power, and the timber he grew to replace coal. This inquiry into the techno-energetic setup of Mehmed Ali's 'industrialisation' sheds new light on the rise of the 'modern' factory as triggered by the Pasha's international trade in agricultural commodities, and how industrialisation under conditions of coal scarcity conditioned Egypt's place in the fossil empire that emerged from early nineteenth-century Britain. Only one type of empire was able to ascend at that moment – fossil-fuelled ones – and any aspirants lacking in the coal endowments required to power these ambitions suffered. This I address in the chapter's conclusion.

Ox Power, Manpower, Horsepower

The Pasha's industrialisation was overwhelmingly powered by oxen attached to rotating wheels that moved machines across various factories. At the height of industrialisation, his factories had no more than eight steam engines, and only one or two of them were functional at all during the 1830s,[8] before the industrial establishment took a downturn in the 1840s. This industrial programme suffered from a scarcity of steam power because coal could only be sourced as an expensive imported commodity, and it suffered as well from a scarcity of educated labourers, technicians, and engineers to work and maintain the engines.[9]

8 For example, Fahmy, *All the Pasha's*, 293–4; Owen, *The Middle East*, 72; Afaf Lutfi Marsot, *Egypt in the Reign of Muhammad Ali*, Cambridge: Cambridge University Press, 1984.

9 An argument made by commentators, then historians, since the 1830s. Auguste Colin, 'Lettres sur l'Égypte: Industrie manufacturière', *Revue des Deux Mondes* 14, 1838,

The distribution and timeline of these few steam engines within the historiography usually come from brief passages in reports by English and French diplomats and commentators from the 1830s. Afaf Lutfi Marsot's survey is representative:

> Muhammad Ali imported the new technology of the power loom, and in 1830, a British engineer set up steam engines for a weaving establishment in Shubra, which had 150 looms, and one in Qalat al-Kabsh with 200 looms. Steam engines were also used in hulling rice in Rosetta and in the foundry at Bulaq.[10]

These engines must have made up the bulk of the eight that powered factories at that point, most of which ceased to be useful by the 1830s. But Marsot's important monograph on Mehmed Ali's reign does not pursue the steam engines much further; nor do other histories.[11]

Substantial parts of the industrial surveys that survived from Mehmed Ali's years were likely based on a book project the Pasha wanted to print in 1834, *Tarikh Rawdat Al-Umran*, or *The History of the Lush Garden of Civilisation*. The project never materialised, but the data collected for the book might have served as a base for European diplomats, including John Bowring, who prepared a detailed report on the status of Egypt in the late 1830s for the British government – still the most cited source on Mehmed Ali's industries.[12] Bowring collected information from British consul general Patrick Campbell, from his own meetings with Mehmed Ali and through his observations from across the country. It may be the case, however, that Bowring's report was compromised by the British foreign minister, Lord Palmerston, who commissioned and edited it. He wanted to see fewer mentions of steam engines and serious industrialisation across its pages. It was produced in the context of the crisis towards the end of the 1830s, when Egypt was coerced by his

517–31; A. E. Crouchley, *The Economic Development of Modern Egypt*, London: Longman, Green and Co., 1938, 72; Ali El-Gritly, *Tarikh Al-sina'a fi Al-nisf Al-awal min Al-qarn Al-tasi'a A'shar [History of Industrialisation in the First Half of the Nineteenth Century]*, Cairo: Dar Al-Ma'arif, 1952, 85; Marsot, *Egypt*, 171.

10 Marsot, *Egypt*, 171.

11 The exception is Ali El-Gritly's volume, which includes a long section on 'motive power' in the Pasha's factories: El-Gritly, *Tarikh*.

12 Ibid., 52–3.

government into opening its borders to free trade, and as such, the question of it producing either agrarian or industrial commodities occupied European commentators.[13]

How were the factories powered? Were there perhaps steam engines that contemporaries and historians failed to account for? The best place to start could be Upper and Middle Egypt, with the sugar factories. This is because the sugar-refining industry had to it a consistency across the 1820s–70s, and it was one of the first industries to turn to steam power, in the late 1840s and early 1850s.[14]

During the 1810s, more than a hundred small sugar and molasses manufactories attested to the long history of small-scale processing of sugar cane in Egypt.[15] In 1818, Mehmed Ali Pasha built the first modern sugar factory. It was set up at Reyremoun, Minya, in Middle Egypt.[16] The factory was managed by an Englishman, a Mr Brinam, and he was succeeded by two Italians.[17] Described in other sources as John Brine from Devonshire, he was probably the first British factory manager hired by Mehmed Ali.[18] The factory ran on 'animal energy', namely oxen.[19] In the late 1830s, Bowring had the opinion that it was 'tolerably well organised'.[20]

Six more factories followed the one at Reyremoun: in the 1830s, Ibrahim Pasha, Mehmed Ali's son, built two, also in Middle Egypt. Four other sugar factories were established in the 1840s, in Middle and Upper Egypt. Two French companies supplied those four factories: Pastré and Cail. Nothing could be found on Pastré's role in the sugar factories of the 1840s, but the Pastré brothers of Marseilles, who moved

13 Marsot, *Egypt*, 244.

14 The later career of the sugarworks (1840s–50s) comes up in Chapters 3 and 4.

15 El-Gritly, *Tarikh*, 59.

16 John Bowring, *Report on Egypt and Candia, Addressed to the Right Hon. Lord Viscount Palmerston, Her Majesty's Principal Secretary of State for Foreign Affairs, etc, etc, etc*, London: W. Clowes and Sons for Her Majesty's Stationery Office, 1840, 22; other sources in Ralph Bodenstein, 'Sugar and Iron: Khedive Ismail's Sugar Factories in Egypt and the Role of French Engineering Companies (1867–1875)', *ABE Journal* 5, 2014.

17 Bowring, *Report*, 22; El-Gritly, *Tarikh*, 59–60.

18 Brine, also known as Brinam in Bowring's report, died at some point in or after 1821. The British travellers heard of his death two months after they met. John Carne, *Letters from the East*, London: Henry Colburn, 1826, 115–16.

19 El-Gritly, *Tarikh*, 60. Bowring, *Report*, 22.

20 Bowring, *Report*, 22.

to Alexandria in Mehmed Ali's time, were active between the 1840s and 1870s in shipping, trade, banking, agricultural and industrial works, and public-works projects.[21] The Pastrés had a partial monopoly on Egypt's western Mediterranean trade during Mehmed Ali's reign, while the other foreign partner in this financing venture of Alexandria, Tossitza, a prominent Greek financier, had the equivalent partial monopoly over the eastern Mediterranean.[22] The second – also French – company that was involved in building these four factories, Cail and Derosne, built many of the modern steam-powered sugar factories under Ismail twenty years later.[23]

Bowring visited the Rhoda factory after 1837. Situated on the Nile, surrounded by 272 acres of sugar cane, the factory had 750 labourers. And there was a very unusual scene waiting for Bowring at the factory:

> Almost without exception, the labourers had mutilated themselves by cutting off the first finger of the right hand, destroying the right eye, or pulling out the front teeth, in order to avoid the conscription.[24]

Historian Ali El-Gritly lists two categories of mutilated persons who were sent by Ali to his factories: those whom the army declined to conscript – after they got themselves mutilated on purpose, seeking this very decision – and those who were mutilated in Ali's wars abroad.[25] In addition to these workers were those who were there as a result of the order issued by the Pasha in 1830 dictating that beggars and homeless orphaned boys be collected and put to work for a daily wage. Another Viceroyal order was issued in 1827 to the governor of Damietta, 'to collect a hundred boys from among those working at tea houses and at tobacco shops' to work as carpenters in the Damietta Arsenal.[26] Convicts were also taken to work at arsenals. Iron and copper

21 Bodenstein, 'Sugar and Iron', endnote 10.

22 Ghislaine Alleaume, 'An Industrial Revolution in Agriculture? Some Observations on the Evolution of Rural Egypt in the Nineteenth Century', in *Agriculture in Egypt: From Pharaonic to Modern Times*, edited by Alan Bowman and Eugene Rogan, Proceedings of the British Academy, Vol. 96, 1999, 336.

23 On the later six factories of Mehmed Ali, see Bodenstein, 'Sugar and Iron'. El-Gritly also for the two factories of Ibrahim Pasha: El-Gritly, *Tarikh*, 59.

24 Bowring, *Report*, 21–2.

25 El-Gritly, *Tarikh*, 110.

26 Ibid., 109–10.

foundries used workers who had been 'caught spinning *barrani* cloth and thread [outside state control]'.[27]

At Rhoda, oxen rotated the wheels that crushed the sugar cane, producing in twenty-four hours four cantars of juice (roughly 180 kilograms).[28] During the two to three months of the sugar-processing season, work went on nonstop. Men took shifts around the clock: 'the works proceed night and day, and when the labourers are weary, and take their rest, they are replaced by others'.[29] A great number of boys and girls were employed.[30]

The Rhoda factory's superintendent provided Bowring with a detailed report of the total expenditures and revenues of the year 1837 that encompassed both factory and plantation. It included no items indicative of the presence of steam power. On things that burn, we have the item 'cost of the firewood' – probably for heating the syrup boilers – but nothing on coal.[31]

Mehmed Ali Pasha invited a sugar refiner from Marseilles (M. Allard) in 1836, and made plans for him to supervise the crafting of 'a steam machine' for sugar refining in the Boulaq foundry, but Allard declined and left, and we have no evidence that such an engine was manufactured in Cairo at that time or later.[32] This affair gives the impression that the Pasha knew that steam technology was key to refining in order to improve his local sugar, with which he wanted to compete with the Marseilles-imported sugar in Cairo. This indicates that it is very likely that only ox power was employed across all sugar factories up to that point.

At the gunpowder workshops in Cairo, the saltpetre was refined first by 'solar heat'. It moved on next to the final stage of chemical processing, operated manually. Bowring visited an extensive workshop, attended by a Frenchman. This workshop had ten mills for pounding and mixing

27 Ibid., 110.

28 Bowring stated that 'cantar' in his report equals 100 lbs, or 45.3 kg. Cantar was an elusive measuring unit of weight, ranging across different places (for example, Turkey, Greece) between dozens of kilograms to over a hundred.

29 Bowring, *Report*, 22.

30 Ibid. Bowring also saw some 'negro children' among the Egyptian labourers. For slavery in nineteenth-century Egypt: Judith Tucker, *Women in Nineteenth-Century Egypt*, Cambridge: Cambridge University Press, 1985, 164–93.

31 Bowring, *Report*, 22–3.

32 Colin, 'Lettres', 528–9.

together sulphur, charcoal, and saltpetre to produce gunpowder. Each of these mills had twenty 'pounders', and each set of twenty pounders was put to motion by a machine moved 'by mule power'. Each mule hooked to each twenty-pounder complex was attended by a labourer.[33] Abdulrahman Al-Raf'i – probably citing contemporary French sources – gave a similar description of the manufactory and said that each of the ten machines moving the pounders was powered by '*Bighal* (mules)'.[34]

Concerning cotton, the ginning – separation of the fabric from the seed – used a simple machine moved by foot, by which 'a workman can separate from 12 to 15lbs per day'.[35] We know that such mechanical installations at this early stage were running on animate energy.[36] Cotton pressing was done also by foot.[37] The pressed cotton bales then moved into factories for further processing into thread and fabric. The biggest establishment concerned with spinning and weaving was the Malta factory in Boulaq, downtown Cairo. It got its name from employing skilled Maltese workers. Here we stop only at the 'spinning' department, not the weaving, confident that the prime movers of weaving looms were the same as those of the spinning jennies, especially given that fully automated power looms were deployed in English factories in the 1840s.

In the late 1830s, the Malta factory had twenty-four carding machines for combing the cotton that went next to the spinning stage, and twenty-eight spinning jennies for spinning the cotton into thread. Each pair of spinning jennies was put to motion by a rotating tambour – a drum or frame for holding the stretched fabric. The factory had fourteen tambours. All the tambours in the factory rotated, and moved the spinning jennies, by a 'machine', powered by eight oxen. Other even bigger spinning factories were to be found in the Delta, and there also, oxen were the exclusive motive power.[38]

But, in some locations at least, the oxen powering the machines were occasionally replaced by men, not steam engines. The Qina manufactory, which employed a thousand people, had its 'moving power' supplied by

33 Bowring, *Report*, 28–9.

34 Abdulrahman Al-Raf'i, *'Asr Mohamed Ali [Mohamed Ali's Reign]*, Cairo: Dar Al-Ma'arif, [1930] 1989, 342.

35 Bowring, *Report*, 20.

36 Alleaume, 'An Industrial', 341.

37 Bowring, *Report*, 20.

38 On Malta/Boulaq and Delta factories: ibid., 33.

100 oxen. But in 'what is called the bersim season, or that of the grass harvest, *the oxen are supplanted by men and boys*.'[39] If he could help it, however, the Pasha preferred oxen to men as motive power. In the middle of the devastating cattle plague of the early 1840s, a superintendent of a factory suggested to the Pasha that an unspecified number of oxen be replaced with 6,500 men in order to power a certain cluster of factories (perhaps Mehmed Ali had asked his men to send suggestions at the time of the crisis). The Pasha, in return, scolded him on two counts. First, he said, 'big minds across the world work ceaselessly on suggesting machines and tools to facilitate labour and save man from types of toil he was not created for . . . [The superintendent] was supposed to follow the lead of those wise men and work on inventions to alleviate the ravages of man, not to extend a suggestion that complicates an issue simplified by wise people.' Second, the Pasha said that if these same 6,500 men were to be put to work farming across the Delta, they would gain the treasury thrice the money that the superintendent suggested would have been saved if they replaced oxen in the factories.[40]

The Malta factory of downtown Cairo was 'the central point of all the rest' of the spinning and weaving business. Spinning jennies and carding machines were manufactured there for use in the new factories in Upper Egypt, and it employed immigrant engineers, technicians, and workers, in all branches related to the textile industry. Bowring did not report any steam engines at Malta. The only prime movers mentioned there were oxen. Nothing across the production chain of cotton-unto-thread and cloth was harnessed to steam engines, even during a time when the Pasha was locally producing spinning and carding machines en masse to be transported to newly established factories. At the moment of writing his report, Bowring knew of the conclusion of construction works of new textile factories across the country, mostly in Middle and Upper Egypt. They were waiting for their machines, which were being manufactured in Cairo.[41]

Factories manufactured machines for mass commodity production that resembled the spinning mills of England – the cover photo of the

39 Ibid., 37. Emphasis added.

40 El-Gritly, *Tarikh*, 87–8, citing correspondences from *Diwan al-Fabriqat* (Manufactories Department), 14 Safar 1257 H [4 April 1841].

41 Bowring, *Report*, 33.

Industrial Revolution if it were a book – and not steam engines. A technical report by an anonymous 'English mechanic' in Egypt looks more closely at this frustration with steam power until the 1830s.[42] The mechanic asked: 'What powers does she [Egypt] possess to impel machinery?' Egypt has no rivers beside the Nile, which did not flow quickly enough to give motive power to mills, or quick streams to use for water power, and it was too late, in 1837, to keep impelling machinery by manual labour: 'in many cases where machines are employed they are worked by the hand'. On the next step up the ladder of prime movers, the English mechanic had in mind horsepower, by which he meant animal power at large. Egypt had some work animals, and imported many, but, the mechanic said, they had better be seen 'in a secondary light'. He returned to the unfortunate lack of water power, while reaffirming that 'water is the chief and most economical manufacturing power in all countries'.[43] This leaves Egypt with one last option, also the most technologically advanced one, the steam engine:

> The only remaining power applicable to machinery is that of steam. Can steam-power be used to advantage in manufacturing in Egypt? Here again Egypt is peculiarly situated; both nature and art appear to be inadequate to supply the means necessary to Egypt's becoming a manufacturing country; she is destitute of all kinds of fuel, coal, wood, or peat; the necessary ingredient in the cost and working of steam-power.[44]

The evidence we present in the next two sections will show that Mehmed Ali Pasha had been acutely aware of these problems since the first day he considered erecting factories. Coal and coal mines are perhaps the keywords most sorely missing from histories of the first half of nineteenth-century Egypt, even though the Pasha ceaselessly strove for these fossil assets as much as he was keen on making and procuring weapons and steam engines.

42 See 'A Report of an English Mechanic on the Manufactures and Agriculture of Egypt', Annex D in Bowring, *Report*, 199–202.

43 Ibid., 199–200.

44 Ibid., 200.

1820s: 'Have you been able to procure coal? Have you bored at all?'

On the origins of the first factories established in the late 1810s, it is often briefly stated that foreign advisors and consuls advised the Pasha to make use of the European factory system. The name of Joseph Bokti (or Bocti), the Swedish consul in Alexandria, usually comes up most prominently among those persons.[45] Expanding on Bokti and Sweden brings some insights on the birth of the Pasha's factories and their strong ties to trade. In Bowring's words, 'the Swedish consul, Bokti, spoke to him [the Pasha] of projects which tended to free Egypt from being dependent upon foreign industry; he immediately put them into execution.'[46] Thedenat-Duvent, the French consul in Alexandria, reported in 1821 that Bokti, the Swedish consul, received in 1819 a substantial sum of money from the Pasha as a token of friendship, and that he was appointed that year the general director of the Pasha's factories.[47] In 1903 *The Journal of the Khedival Agricultural Society* published an essay that captured the official story on Bokti well into the twentieth century:

> It was Mr. Bokti the Swedish consul who about 1814 advised Mohamed Aly to start manufactories in Egypt. The machines were ordered from France and England, and European mechanics were engaged to teach the natives.[48]

Joseph Bokti, however, belonged to the world of Syrian merchants and trade more than he belonged to Sweden and diplomacy, by far. Joseph's father and uncles, all Syrian Christian traders, moved to Egypt in the eighteenth century. They took to different cities in the usual manner of commercial households of the time. Some went to Cairo, others settled in Alexandria and Damietta, while one of the brothers relocated to Livorno. Joseph Bokti, after the first generation, established

45 See, for example, Crouchley, *The Economic*, 67.

46 Bowring, *Report*, 31.

47 R. P. Thedenat-Duvent, *Mehemed Ali, Vice-Roi D'Egypte*, unknown publisher, 1821, 90.

48 Judge Sandars, 'The Cotton Plant', *The Journal of the Khedival Agricultural Society and the School of Agriculture* 5:6, 1903, 272.

strong trade interests between Europe and the Levant, and was hired by Sweden as its consul, while another Bokti became the Russian consul.[49] The Syrian Christian merchants dominated the eastern Mediterranean markets, including imports of European textiles and the reselling of cloth. As such, his career can be seen as a vital node of trade between Sweden, the Mediterranean, and Egypt, where the Pasha sought his services. Bokti was involved in Egypt's trade in grain in exchange for Swedish pig-iron, and Swedish ships were chartered for selling the Egyptian grains across the Mediterranean. Given the imported iron and Bokti's involvement in the textile trade, it might have come naturally for him to suggest cotton industries and factories, and the building of the silk factory at Khurunfish, Cairo.[50]

Also significant in regards to the earliest development of factories and industrialisation under Mehmed Ali is the fact that Egyptian-British commercial links prior to the 1820s did not focus on manufactured commodities. It was the trade in grains – mainly for the British troops across the Mediterranean – that coloured these exchanges in the 1810s. Only after 1820 did British trade have a major stake in Egyptian long staple cotton.[51]

It seems that trade and commercial capital motivated and inspired industrialisation that was to produce more commodities for further trade. This does not suggest that we should downplay the role of the Pasha's army in inspiring industrialisation – it was the greater domestic consumer of the factories' output. However, this Swedish-Syrian detour suggests that industrialisation originated for multiple reasons, with consequences that reinforced both (that is, the military and industrialisation itself), through a series of ad hoc situations as opposed to a thoroughly planned programme for feeding military expansion by building factories. It is tempting, under this light, to see the army build-up and

49 Marsot, *Egypt*, 167. On the Bokti brother hired as consul for Russia: Thomas Philip, *The Syrians in Egypt, 1725–1975*, Stuttgart: Steiner-Verlag, 1985, 65. Peter Hill recently brought up the tradition of hiring Syrian merchants to European consular positions: Peter Hill, *Prophet of Reason: Science, Religion and the Origins of the Modern Middle East*, London: Oneworld Publications, 2024.

50 Marsot, *Egypt*, 65. Fortunately, we have a detailed study based on rarely accessed and rich archival material (*Diwan Al-Fabriqat*) on the labour conditions and composition of the Khurunfish silk factory; see Ghazaleh, 'Manufacturing', 124–35.

51 Jonathan Parry, *The Promised Lands: The British and the Ottoman Middle East*, Princeton and Oxford: Princeton University Press, 2022, 147–8.

military expansion across the 1820s and 1830s as growing in tandem with expansion in industrialisation: more factories producing more commodities for growing trade, making it necessary to develop the military to protect the growing commercial empire and bring in necessary raw materials for industrialisation, like iron, copper, and coal; then, with the army expansion, local demand for certain commodities – textiles, ships, cannons – grows, pushing for more expansion on the industrial front. Growth begetting growth, in other words, where trade in commodities moves the cycle.

Pig-iron was not enough on its own to make machines. Systemic British and French involvement in building the Pasha's factories and supplying them with machines and some engines starts with a lead from the British House of Commons.

In the spring of 1824, British laws were still banning exports of machinery and the labour of machine artisans. This was the reason why the earliest modern machinery Mehmed Ali brought to Egypt was from France, not England. The records of a House of Commons committee on the matter show evidence that several British capitalists had, in the early 1820s, established in France substantial machine and engine-making workshops that relied on smuggled British iron ore and on artisans who smuggled themselves from England. Thomas Lester, a witness brought before the Select Committee on Artisans and Machinery, was, in his words, in the business of 'the putting of engines together'. From Tipton, Staffordshire, he travelled for a year to work for a Mr Manby in France, where he assembled three steam engines for steamboats. The steam engines workshop in Charenton, outside Paris, hired around 200 Englishmen and boys, mostly from Staffordshire. The other (unskilled) workers on the floor were French. Lester claimed that Mr Manby held the artisans' passports to keep them from going back to England.[52] Generally, E. P. Thompson gathered that, around that year, 'agents of foreign employers scoured Britain in the hope of enticing skilled men [in machine making] to France, Russia, Germany, [and] America'.[53]

52 Excerpts from *Third Report from Select Committee on Artizans and Machinery*, House of Commons, including the questioning of artisan Thomas Lester, in 'Third Report', *The British Press* [a newspaper], 9 April 1824.

53 E. P. Thompson, *The Making of the English Working Class*, New York: Vintage Books, 1966, 246.

Meanwhile, the committee's investigations suggested that the Pasha of Egypt, in or just before early 1824 – a few years after Bokti and others had advised on factories – 'applied in England for cotton machinery, because he was going to grow cotton; upon which he received an answer from different machine makers in England, through his agent, that he could not get the machinery in England, because it was prohibited'.[54] The Pasha directed his inquiries for cotton machinery to France, and a workshop wrote back to Cairo that the required machines would be supplied at inflated prices. Steam-powered machinery was being made in France at the time for buyers in Switzerland, Germany, and Belgium.[55] Shortly after this leakage of steam technology's know-how, the prohibition on experts' and engines' movements, at least in the case of Egypt, was repealed.

Thomas Galloway, an engineer, travelled to Egypt in 1825 to erect a copper mill and to provide machines for dredging irrigation canals, and then retained for many years the position of chief engineer of Mehmed Ali.[56] He provided (through correspondences exchanged with his father, Alexander Galloway, the machine-builder and one of London's biggest employers, and his brother John) rich evidence over that year and the next on the earliest systematic British involvement in mechanising Egypt. His father, whom the labourers at his London workshop called Alex, was, around the time of the French Revolution, a radical Spencean (a movement that advocated the destruction of private property) and a leading figure in the notorious British-Jacobin London Corresponding Society (LCS), which the government disbanded, imprisoning many of its members, including Alexander. He believed, along with LCS members, in universal voting rights for all men, and led a campaign, in

54 Select Committee on Artizans and Machinery proceedings, questioning of a Mr Alexander, reproduced in 'Actual State and Condition of the Manufacturers in France', *Bell's Weekly Messenger*, 19 April 1824.

55 Ibid.

56 Dredging machines and a copper mill came up in Thomas Galloway's letters to his father: 'Improvements in Egypt', *The Scotsman*, 19 October 1825. A Mr Galloway coming to Egypt to establish a copper mill is also mentioned in Mehmed Ali's orders. See Rauf Abbas (Editor), *Al-Awamir wa Al-Mukatabat Al-Sadira Min 'Aziz Misr Mohamed Ali [Orders and Correspondences Issued by the Wali of Egypt Mohamed Ali] Vol. I*, Cairo: Dar Al-Kutub, 2010, 231, item 1105 [order dated 18 Rabi' Awal 1244 H [27 September 1828].

the 1810s, for enhancing payment terms for artisan apprentices.[57] Whether ideological or business-expansion motives were behind this Galloway's move to Mehmed Ali's Egypt remains a mystery.

In March 1825, Thomas Galloway landed in Alexandria, becoming most likely the first representative of a British steam engines and machinery manufacturer to visit Egypt after the artisans and machinery exportation prohibition was reversed. Several British newspapers copied the description of his first batch of letters sent home to his father,[58] speaking of how the Pasha had that far improvised and laboured towards gaining some of Britain's machines and mechanical knowledge, and of an upcoming monumental shift to mechanical industrialisation on the Nile:

> It will be seen [from the Galloway letters] that the Pacha has contrived to possess himself of the last and highest improvements in our manufacturing machinery. [. . .] We find him [i.e. Mehmed Ali] surrounded with the intrigues of European adventurers, and using the most dexterous policy to obtain the most profitable information from them. He has his choice, and apparently he is quite equal to the task of selecting, from all the superb inventions of modern engineers, French or English. The Earl of Liverpool has declared that the country had been raised to its present greatness by Arkwright, Bolton, Watt, and others. *Egypt now comes in for the largest share of the advantage.* With a soil of the richest and cheapest abundance in the production of grain, she begins her career in manufactures, and starts from the advanced point of mechanical knowledge, attained by the philosophy of all Christendom.[59]

57 E. P. Thompson estimates that Alexander Galloway was undoubtedly a Spencean, a radical movement that sought to undermine aristocracy, including through the destruction of 'private property in land'. Thompson, *The Making*, 161. See also on Alexander Galloway's lingering radicalism in the early 1820s: Nile Green, *The Love of Strangers: What Six Muslim Students Learned in Jane Austen's London*, Princeton and Oxford: Princeton University Press, 2015, 264–5.

58 Source unidentified, but apparently the family found the letters important enough to share with the press.

59 'Improvements in Egypt', *London Packet and New Lloyd's Evening Post*, 17 October 1825. Emphasis added. Shorter version in 'Egypt', *Manchester Guardian*, 22 October 1825.

Mr Galloway found out, once he landed, that the country possessed a 'great quantity of cotton machinery', which 'has, first and last, been sent here from France, and that for no other reason, as they avow, but because it could not be had from England'.[60] However, most of the next wave of textile-related machinery was manufactured in Cairo, not London or Paris, under French supervision, using the designs and the expertise already in place.

It is likely that, at least since the year 1819, Mehmed Ali was aware of the necessity of steam engines, and was no less aware of the fact that they needed cheap and abundant locally acquired coal to operate. The high price of imported coal did bite hard, driving this fact home. In 1823, one of his orders reveals that, due to the 'grave expenses of the coal imported for the manufactories, [the Pasha says] perhaps cotton wood and cotton seed can be burned in its stead', and he ordered experiments to this effect in the iron foundry and the guns workshops.[61] Perhaps at this point (on or before 1823) he had imported from the French workshops a few steam engines, or perhaps his search for alternatives to coal came about as the alternative fuels were to be used in gunpowder at the guns workshops and direct heat energy applied to casting iron at the foundry. Information from this point in time is almost totally absent, and ambiguity is compounded by the fact that words such as cotton 'machinery' might or might not have referred to machinery compelled by steam engines. However, through his several meetings with, and machine requests from, Thomas Galloway in 1825, we can see that Mehmed Ali saw steam power and abundant fossil energy as inseparable. Simultaneously, he was powering his burgeoning industry with animal power and requesting specific machinery for the purpose – that is, until he found coal mines to feed steam power. The young Galloway told his father in a letter dated 30 July 1825:

> I have been with the Pacha this morning, during which time we had a long conversation on the subject of steam, with which he appears much pleased. There will be a fine opening here for steam-engines in general. There has been an immense coal mine discovered near

60 Ibid.

61 Order dated 11 Muharram 1239 H [16 September 1823], in Abbas (Editor), *Al-Awamir Vol. I*, 97, item 466.

> Constantinople, and the Pacha expects to gain permission to work it from the Grand Seignor [i.e. the Ottoman Sultan]. I have offered to take the superintendence of it.[62]

The letter does not reveal whether the Pasha told Thomas about the coal mine near Constantinople, or whether Thomas had learned about it at an earlier time. But, since 1820 at least, the Pasha was looking vigorously for coal. Meanwhile, until he was to find his mines, he requested from the Galloways machines that could be powered by cattle.

Coal (that is, its absence) and cattle-mills were the problem and provisional solution that characterised Mehmed Ali's industrialisation well into the 1840s.

Concerning coal, in September 1819, the Pasha declared 'a very great reward to any that shall discover a coal mine in Upper Egypt' after news came that year of iron and lead mines discovered there.[63] In October 1822, the Pasha ordered his military commander in Sudan to assist and protect a certain 'M. Rubil [name transliterated from Arabic] and two other persons with him, who are coming to Upper Egypt and the Sudanese districts to search for minerals'.[64] His hopes to find minerals in Sudan, probably including coal, were a substantial factor for deciding to send military expeditions. This is based on the Pasha's own words. On 29 October 1821, he congratulated his military commander in Sudan for 'subduing Kordofan [. . .], and [declares that] the reason for sacrificing money and lives and facing great hardships, is *only* to bring slaves in great numbers *and* [. . .] search for places of minerals and such things. [The Pasha] calls on him to persevere in his efforts to that end.'[65] The next order, on the same day, was to inform another commander in Sudan that some metallurgists were on their way from Cairo 'to search for sites of minerals'.[66]

On or before April 1824, Mr Wilson, an English traveller to Egypt, said that, generally, 'coal mines have been discovered [which later

62 'Improvements in Egypt', *London Packet and New Lloyd's Evening Post*, 17 October 1825.

63 'Egypt – Canal of Rosetta', *The Scots Magazine*, 1 May 1820, 458.

64 Order to the military commander in Sudan, dated 15 Muharram 1238 H [1 October 1822], in Abbas (Editor), *Al-Awamir Vol. I*, 70, item 320.

65 Order to Mohamed Bey Al-Daftardar in Sudan, dated 3 Safar 1237 H [29 October 1821] in ibid., 46, item 186. Emphasis added.

66 Order to the military commander in Sudan, same day, in ibid., 46, item 187.

evidence discredits]', and English mineralogists were supposedly employed to search for mines in Egypt.[67] The expedition that established Egyptian rule in Sudan was reportedly motivated by the search for gold to finance commercial expansion and for slaves to build a new army.[68] However, further evidence from the 1820s and 1830s, given shortly, and in the next chapter, suggests that his correspondences and orders about minerals in Sudan likely included an interest in finding coal seams.

In 1824, an unnamed traveller met in Alexandria a Mr James Burton, whom the Pasha hired 'as a mineral observer'. He had found for him copper only, seemingly in abundance, between Suez and Qusair along the Red Sea mountains, but there were also from Burton's researches 'indications of coal towards Syria'. Next, the mineralogist told our traveller that 'coal was the great object of the Pacha, as the want of fuel prevents him profiting by the chemical productions which the country affords [i.e. working discovered minerals into machines and weapons]'.[69] When Thomas Galloway landed in Egypt in 1825, he learned of another British expert, Mr Brown, who was hired by the Pasha – perhaps after Burton – to inspect the iron, lead, and copper mines, 'and if he finds a sufficiency of fuel to work these ores', the Pasha was to start getting his local supply of such minerals.[70] In July 1830, a Swedish metallurgist, Mr Ginsburg, was to travel to Suez to search for coal. The governor of Suez was ordered to assist him with his travels.[71] His search seemingly kept him engaged for at least the next three years.[72] The great coal search went on for years; in fact, it extended across Mehmed Ali's reign, and featured strongly in his adventures in Syria in the 1830s (as we shall see in the next chapter).

Concerning the oxen-powered machines the Pasha requested from the Galloways, we find several pieces of information in Thomas's letter

67 'The Pacha of Egypt', *Dublin Evening Mail*, 30 April 1824.

68 Marsot, *Egypt*, 205; Fahmy, *All the Pasha's*, 40, 86–7.

69 'Egypt', *Caledonian Mercury*, 19 July 1824. James Burton was also involved in excavations for ancient Egyptian artefacts: Parry, *The Promised*, 154.

70 'Improvements in Egypt', *London Packet and New Lloyd's Evening Post*, 17 October 1825.

71 Order to head of Khedival Diwan, 1 Safar 1246 H [22 July 1830], in Abbas (Editor), *Al-Awamir Vol. I*, 299, item 1427.

72 Order to Habib Affandi dated 3 Rabi' Awal 1249 H [20 July 1833], in ibid., 371, item 1815.

to his father and brother. The Egyptian shopping list sent to London included a small-scale paper mill. Although it was planned to be erected on the Nile, 'the machinery must be driven by cattle, as there is no possibility of applying water power, except by artificial methods', Thomas told his brother.[73] Implicitly then, a steam engine for the mill was also out of the question, but it was an engine nonetheless, redesigned to be powered by oxen. On his first tour of the Malta textile factory and foundry of textile machines in Boulaq, where he observed machine manufacturing that used the designs of the first French textile machines imported earlier as the reference, Mr Galloway saw forty turning-lathes for manufacturing iron rollers to be fitted to textile machinery, all powered by cattle. At the iron forges, there were 'large bellows driven by cattle', for heating the iron worked into machine parts.[74] The piece demanded from London to power the paper mill was to fit into this arrangement.

A letter that Thomas got from his brother John the next year included mentions, designs, and sketches of several animal-powered engines from the London Galloway workshops. For example, attached to the letter were designs of machines and implements of different uses – lathes, saw-mills, wheel-engines, grindstones – 'driven by a six-horse engine', which was also amenable for powering 'a most formidable jenny'. What motive power was in the range of a 'six-horse engine' in Egypt at the time or slightly after? The eight-oxen-powered tambour that moved several sets of jennies Mr Bowring described in the 1830s. Perhaps the most formidable jenny was deployed, being the tambour with eight oxen. John Galloway had also sent separately sketches of 'two cattle-mills' that Thomas had earlier requested.[75]

Beside the machinery and animal-powered engines that Thomas requested, John sent an unsolicited 'set of drawings for a large foundry, with steam-engines, blowing-cylinders, furnaces, etc.', with the view that if the Pasha wanted it, it was to cost £5,000–7,000. The motive for his labour on these designs was that he heard of a call from the Pasha in London for an iron furnace. He did also express his concern over motive

73 'Improvements in Egypt', *London Packet and New Lloyd's Evening Post*, 17 October 1825.

74 Ibid.

75 All citations in paragraph: letter dated 10 October 1826, partially reproduced in 'The Galloway Correspondence, from the *New Times*', *London Courier and Evening Gazette*, 26 June 1827.

power, if the Galloway workshops of London were to expand their business with Egypt. Coal mines and water power occupied John Galloway greatly. He asked his brother in Cairo:

> The great drawback to the manufacture of Egypt is the want of a moving power. Have you been able to procure coal? Have you bored at all? This is a point you ought to press the Pacha to persevere with. Have you ever seen any water falls? Do they apply water-wheels of any, and what sort? Do they make use of the cataracts in Upper Egypt? It appears to me that machinery will not be proceeded in with vigour unless coal can be procured, or the water applied for the working of water-wheels, as all cattle machines are bad on account of their irregular motions. You must direct your attention to this. I am of the opinion that waterpower may be used more than it is.[76]

John also advised his brother, perhaps hopeful that mines or water power would come to the rescue, to get as much information as he could on sugar cane in Egypt, with the view that they could convince the Pasha to buy sugar mills. With a reassuring note, he added, 'Now, get your credentials, money and orders, and then we can shew the Egyptians what can be done for them. [. . .] all you can order, we make, and only shew us that more is required, and it can be done.'[77]

Wind, Water, and Trees

It seems, however, that John Galloway's ambitions for mechanising Egypt stayed – like those of Mehmed Ali – besieged for long by her coal-free geology. This frustrated steaming of industrialisation came combined with efforts to explore and work towards alternative prime movers and combustibles. We turn to this next. Wind, water, and trees take the story forward.

76 Ibid.

77 Ibid. Notably, John followed his brother Thomas Galloway Bey to Egypt at some point. In 1837, their father said at a House of Commons summons that both his sons were in Egypt, for years. House of Commons, *Report*, 65.

Starting with wind, in the late eighteenth century, harnessing its power was probably confined to Alexandria, where very few windmills were brought by individuals from Europe.[78] In the early 1830s or earlier, the Pasha erected thirty windmills there.[79] Their construction took so long, and they might have faced other complications in their operation, that he decided to give up on this energy mode, and replace the wind with cattle to run these mills.[80] Perhaps the strongest presence of wind power across the nineteenth century in Egypt was in the mid-1850s, under Sa'ed Pasha. Around 1855, Nassau William Senior, a lawyer and economist, saw along the coast two miles from Alexandria 'a forest of windmills' used for grinding wheat for the Egyptian army. These windmills might have been the same that were erected then abandoned before operating, given Mehmed Ali's frustration with the pace of constructions, in the 1830s.[81]

Mehmed Ali hung some hope on water power. By the 1830s, when many of the biggest irrigation projects were concluded, the Pasha wanted to harness the power of the few small artificial waterfalls of the waterworks. At the Zaqaziq barrages, in 1836, he ordered three Egyptian engineers to build a rice-hulling establishment, 'and to provide a report about the months in which water currents are strong enough to move the wheels of the mill by the water cascading from the barrages' fall'.[82] The same barrages apparently were mentioned a few months earlier in combination with plans for a paper mill in Al-Ja'fariya, Sharqiya.[83] Generally, Linant Bey, minister of public works for Mehmed Ali, thought that the artificial waterfalls caused by barrages on major canals in the Delta, and those of a planned one – never realised – at Gabal Silsila in northern Aswan, were potential sites for establishing water-powered factories.[84]

78 Rauf Abbas, *Al-Haraka Al-'Umaliyya Fi Misr [Labour Movements in Egypt] (1899–1952)*, Cairo: Supreme Council of Culture, [1967] 2016, 40.

79 El-Gritly, *Tarikh*, 85.

80 Ibid., 86.

81 Nassau William Senior, *Conversations and Journals in Egypt and Malta, Vol. I*, London: Sampson Low, Martson, Searle & Rivington, 1882, 12.

82 Viceroyal order dated 10 Muharram 1252 H [26 April 1836], copied in Amin Sami, *Taqwim Al-Nil 2 [Survey of the Nile Vol. 2]*, Cairo: Matba'at Dar Al Kutub, 1936, 465.

83 El-Gritly, *Tarikh*, 85.

84 Linant de Bellefonds Bey, *Memoirs sur les Principaux Travaux d'Utilite Public Executes en Egypte Depuis la Plus Haute Antiquite Jusqu'a nos Jours*, Paris: Arthus Bertrand Editeur, 1872, 427.

The grandest plan for water-powered spinning mills was drawn in 1843, but to what extent the project was serious remains unclear. That year, Egypt was in the middle of a disastrous cattle plague that devastated cattle power at the factories (see next chapter), which might have motivated this project, or generally catalysed thinking of alternatives to the cattle in the factories. The only eyewitness to the affair never brought this up. Engineer Linant Bey, in his memoirs, hovers around two other motives. First, he thought that the Pasha was at the time disillusioned with the dismal state of the machinery in the spinning mills, with all the dust and humidity affecting the performance of the delicate spinning machinery, and the deteriorating state of supervision and maintenance. Second, he hinted at this being just a political manoeuvre by the Pasha to placate his governor in Sudan, who was on the verge of open revolt. Linant subtly hinted at the Pasha using the whole affair as part of a plot to assassinate the rebellious governor. At a meeting with the French consul in September 1843 in Cairo, Mehmed Ali reportedly said that he was going to use the services of M. Linant in providing water power at the cataracts of Aswan for a large complex of cotton mills, then gift it all to the governor of Sudan, who was to come and meet him at Aswan, choose the location of the mills himself, and have it built and gifted to him and his family, and thereby become the happiest of his governors.[85]

On the road to Aswan with Linant, the Pasha gave orders to district governors along the way to procure and send to Aswan bricks, lime, and other building material. Given that he took the trip with the engineer to do preliminary investigations on the feasibility of the project, these orders for building material then might have been his way of spreading the news that he was serious about it, which supports the hypothesis of the political manoeuvre that targeted his angry Sudan governor. Linant's studies proved that it was technically possible to harness water power for the mills, with the help of a dam, canal, and sluice systems, but he also reminded the Pasha that the site was too far from both cotton production and population centres that were to supply hands. Linant concluded his account of the unrealised Aswan cataracts water power project by

85 Ibid., 424–7.

saying that the governor of Sudan died of fever even before Mehmed Ali was back in Cairo from Aswan.[86]

With or before the alternative prime movers of wind and water power, the Pasha invested heavily in unfossilised alternatives to coal. There is evidence that there were at least two kinds of trees producing exceptionally useful charcoal: one, the *sant* tree (*Acacia nilotica)*, for charcoal used as fuel for motive power; the other, the *sefsaf* (willow) tree, whose wood-turned-charcoal was used for making gunpowder. These two applications of wood, as fodder for both steam engines and guns, the two great makers of empires, won the Pasha's attention in the late 1820s. Between July 1828 and June 1830, Mehmed Ali grew in the Delta alone 4.3 million trees, 1.1 million of which were *sant*, and 300,000 were *sefsaf* trees.[87]

The career of the *sant* trees across the early 1830s, when they started to grow, seems closely connected with producing charcoal-as-fuel. The same applies for *sefsaf* trees and charcoal-as-gunpowder. Here, I explore briefly how the evidence we have on the two trees tilts towards showing that the Pasha started this project systematically at scale in the late 1820s.

On 27 February 1831, Mehmed Ali ordered the growing of *sant*, *labkh*, and other trees on 700 acres in Tanta, Gharbiya. The intention was to grow the trees intensively like a forest, irrigated with water from the Nile. The 700 acres were chosen carefully earlier by 'his representative', and the order included instructions on providing 'enough *saqiyas* [irrigation waterwheels], [. . .] men, cattle, and gardeners' for the project.[88] These 700 acres might have been the latest chunk in a bigger batch of land dedicated to *sant* forests in Gharbiya. In late May 1831, another Viceroyal order revealed that, at that point, 'because of the necessity of *sant* wood for making coal', at least 6,000 acres were under *sant* trees in Gharbiya (perhaps these 6,000 acres hosted some of the 1.1 million *sant* trees from the turn of the decade), and that 'one or two thousand acres should be used [to cultivate *sant*] in Upper Egypt, and

86 Ibid. Ali El-Gritly stressed technical aspects (absence of dust, dry weather, and so on) in his account of the Aswan plans, but not the political aspects that Linant reported. See El-Gritly, *Tarikh*, 64–5.

87 Sami, *Taqwim Vol. 2*, 379, probably citing *Waqa 'i' Masriya* of Ramadan 1246 H, since the pieces of information before and after this table are cited from it.

88 Order to the *Ma'moor* of Tanta, dated 15 Ramadan 1246 H [27 February 1831], in Abbas (Editor), *Al-Awamir Vol. I*, 321, item 1536.

the cost will be shouldered by the treasury'.[89] If the thousands of acres across Gharbiya and Upper Egypt used for growing *sant* trees followed the same intensive production pattern of the 700 acres in Tanta, it must have been a major attempt to produce alternative fuel – charcoal – economically.

It might have been the case, however, that the Pasha was unaware until then of the extensive technical expertise required for turning *sant* wood, efficiently and economically, into charcoal.[90] On 10 August 1835, after he witnessed the making of charcoal from *sant* in Upper Egypt, he asked the governors of the Delta provinces:

> Did you not see the likes of this process in your districts? Are there no artisans for this craft in Qalioubiya for instance? Tell me what you know. If such artisans are absent [from your districts] they must be brought from Upper Egypt, because this fulfils two benefits, one for the people and the other for the *miri* [state]. Make coal [*sic*] and do not deprive the people of this blessing.[91]

The next week he ordered one of the two governors of Upper Egypt

> to send several individuals [experts in this art] to the district of Munufiya to teach its people. Give these individuals the salary of several months instantly. After they teach the people of Munufiya this craft, they are to return to their district, and some of them will later be dispatched as teachers to the Delta districts.[92]

Concerning *sefsaf* trees, they were again requested to be increased a few years after the 300,000 trees-batch was planted in 1828–30. In March 1833, an order to all district administrators included that every person

89 Order to Upper Egypt's director, dated 18 Zu al-Hijja 1246 H [30 May 1831], in ibid., 326, item 1567.

90 At least since the mid-eighteenth century, Dandara in Upper Egypt was growing trees and manufacturing charcoal that covered most demand in Egypt. Fred H. Lawson, 'Rural Revolt and Provincial Society in Egypt, 1820–1824', *International Journal for Middle East Studies* 13:2, 1981, 131–53, 137.

91 Order to the governors of the Delta, except Giza and Behira, dated 16 Rabi' Thani 1251 H [10 August 1835], in Abbas (Editor), *Al-Awamir Vol. II*, 9–10, item 2239.

92 Order to the governor of half of Upper Egypt, 27 Rabi' Thani 1251 H [21 August 1835], in ibid., 12, item 2249.

planting such a tree shall get 'some money' as incentive, and all trees and wood harvested 'shall be sent quickly to the gunpowder factories'.[93] A tree discovered in Sudan, said to have been used to make gunpowder, was reported as 'producing gunpowder of a quality higher than that made from *willows*'.[94] Willows were apparently grown on scale to feed, as charcoal, the making of gunpowder.

Conclusion

This chapter departed from the conundrum of how Mehmed Ali was – and was not – the founder of modern Egypt. Modernity, in this particular historical sense, understandably covers much more than modern fossil-fuelled factories. Explicit and implicit in associating the Pasha with modernity in Egypt are his relentless endeavours towards radical overhauls in education (initiating secular schools and educating Egyptian technocrats and engineers in Europe), agrarian production (perennial irrigation and associated infrastructures, and developing monocrops for trade), military organisation (building an aggressive modern army), and state administration (centralising and organising administration, initiating processes of censuses and cadastral maps). Coming in tandem with the rise of fossil-fuelled industrialisation and the steam engine out of Britain at the time, Mehmed Ali's radically expanded factories programme has captured historians' and popular imagination alike in terms of thinking of this modernity. Possibly, it is assumed that these factories were modern simply because they were built at a time of many local endeavours towards modernity and because this was the epoch of the modern factory and its machines, even while we know from the history that their motive power was not itself modern. The sheer output of several industries – for a while – likely entrenched this blind spot on the composition of the factories during the early nineteenth-century Egyptian industrialisation.

Exploring this contradiction, initially just to reveal this blind spot, I investigated the animal and human motive power across several

93 Order to council director dated 3 Zulqi'da 1248 H [23 March 1833], in Abbas (editor), *Al-Awamir Vol. I*, 362, item 1763.

94 *Waqa 'i' Masriya*, 12 Rabi' Thani 1249 H [26 October 1833]. Emphasis added.

industries during the high point of this industrialisation, the 1830s: sugar mass-produced on scale in factories powered by oxen. Gunpowder for the regionally aggressive army was produced in mills run by mules, with impressive results. Cotton pressing, spinning, and weaving were conducted on a likewise impressive scale, by means of machines whose blueprints were copied from British and French fossil-fuelled technology and built locally in foundries with engines that were hooked to teams of eight oxen, potentially the same six-horsepower engines designed by the Galloways. The foundries themselves also produced cannons for the army and the navy and were, again, powered by oxen and men.

Now, from this description of the motive power in the factories, we can remove from Mehmed Ali's modernity – for good – the concept of modern industrial production. His industries were based on the most traditional kinds of prime movers. More importantly, for the purposes of this book, this investigation of the energy-technology setup of industrialisation gives some leads on the early moments of the Capitalocene in Egypt and across the globe. Let us conclude with three of them.

First, Egypt's first steam engines – imported from France, because England embargoed them until the mid-1820s – were acquired with the intention of powering factories to expand Mehmed Ali's international trade. The first person on record to advise the Pasha to industrialise (Joseph Bokti, the Syrian merchant/Swedish consul) represented a well-established mercantile family from the Levant and was heavily involved in the Pasha's trade in grain and iron with Europe. It is plausible to argue, then, that the manner in which Egypt was connected to the budding industrial capitalist core at the time – via trade in raw materials – was what inspired industrialisation. That it was the origination of the idea itself, as opposed to a thorough plan to industrialise and civilise Egypt, at a time when fossil capital was concentrated in Britain, driving up demand for raw materials to produce industrial commodities. First, you monopolise agrarian commodities and trade in them with Europe, then someone tells you that adding factories to the production base for raw materials already established – to further process these commodities – will bring even more profits through more trade. This fine-grained argument seeks to highlight certain dynamics involved in how this inclination towards expanded international trade, connected to coal and steam engines in Britain and eventually western Europe – whence the factories started to increase demand for raw materials from Egypt – inspired Egypt's turn to industrialisation. In the case of cotton, its expanded production gave rise to the idea that the commodity would bring more profits if the Pasha processed it in factories into thread and cloth, even if he started growing it for exporting and overhauled agrarian infrastructure towards perennial irrigation for the purpose. The whole process of industrialisation for trade might have initiated the creation of a modern army to protect it by creating a regional empire, which was simultaneously sent shopping for industrial production's inputs, slaves and minerals, including coal, in Sudan, and later in Syria (1830s; see next chapter), as much as the army later consumed the commodities produced through industrialisation. Then, finally, one overarching note tying what was happening in Egypt to the Industrial Revolution is that the factor that unleashed all of this was perhaps something above and external to these dynamics: namely, the process of accelerated commodity production made possible by mechanised fossil-fed machinery emerging at the time out of England.

Second, when these industrialisation plans predicated on trade-oriented production took form, native fossil capital, and industrialisation proper, could not materialise because Egypt lacked the cheap coal that made it possible in Britain. Coal's absence conditioned Mehmed Ali's industrialisation and kept it, until its demise, transitional, inherently in crisis, perpetually waiting for fossil-fuelled motive power. This is evident in the plans of the Galloways for Egypt, their intensive preoccupation with boring for coal, and in some of the machines they designed for the Pasha's factories to accommodate ox power instead of coal-fuelled steam engines until boring for coal brought solid results. It is evident in Mehmed Ali's relentless but fruitless search for coal mines across the 1820s, and in frustrated parallel plans for alternative motive powers, wind and water. It is, finally, evident in the extensive scale of growing and irrigating planned forests in the Delta and Upper Egypt – alongside impressive fields of cotton, grains, and sugar cane – to produce charcoal, in an attempt to reproduce the subterranean foundation of Britain's industrialisation on Egyptian soil. If only coal seams were to be uncovered, if only shafts were dug and peasants were sent down to bring up coal, a fossil empire on the Nile would have become a possibility.

Fossil capital's absence is what frustrated Mehmed Ali's endeavours to emerge as what, in hindsight, we can describe as a world-class industrialist independent of England with his own guns and steam engines. Access to coal in abundant quantities was a foundation for the most advanced industrial-capitalist country at the time, Britain, but not to Mehmed Ali's Egypt, where his ambitions did not find matching carboniferous resources.

Finally, returning to the contradiction we began with – the Pasha's impressive commodities output volume from ox-powered industrialisation – the overview is of the aborted attempt at the making of a grand and terrible fossil capitalist (Mehmed Ali). He created a system that was able to function for decades, waiting for the crucial missing ingredient of coal, which he kept searching for to ignite steam power and, in the process, also ignite the Egyptian part of the fire of large-scale fossil fuel combustion. This system, given the absence of steam power that could have provided a huge amount of labour hours and freed many hands that held the *shaduf* for irrigation on the field and the whip for the oxen in the factories to do his bidding elsewhere, was a highly predatory state, as indicated by long-established historical evidence on the Pasha's reign. But adding the techno-energetic aspects of this state to what we already know offers us openings for rearranging causalities – and reading contingencies – of his reign, in the global context of the then-emerging Capitalocene.

The early nineteenth-century Egyptian Pashalik dreaming of limitless supplies of coal to ignite steam engines of industry and war, more profitable agrarian and industrial production and trade, and a far more powerful military to protect it from coal-rich Britain ran into a protracted power crisis in the 1830s and 1840s. The crisis upended the fossil-empire dreams for good and initiated long decades of ever-growing dependency, and then military occupation, towards the end of the century. To this we turn in the following chapter.

2

The Power Crisis (1838–48)

A most remarkable development from the 1840s in Egypt was how landownership patterns witnessed radical change, towards larger and greater concentration of land in the hands of fewer proprietors: the origination of the Egyptian nobility class of the nineteenth century.[1] The end of Mehmed Ali's monopoly system, largely the result of a European-brokered treaty for free trade in the Ottoman realm in the wake of the Pasha's Syrian wars, and the ultimate failure of his industrialisation programme by the early 1840s are two developments intimately connected with this epochal land redistribution.

By revisiting the 1830s–40s transformations in landownership, together with the Pasha's Syrian wars of the 1830s, this chapter takes the analysis on the techno-energetic developments in Mehmed Ali's Egypt

1 On landownership in nineteenth-century Egypt, with a focus on the 1840s: Roger Owen, *Cotton and the Egyptian Economy, 1820–1914*, Oxford: The Clarendon Press, 1969; Alan Richards, 'Primitive Accumulation in Egypt, 1798–1882', *Review (Fernand Braudel Center)* 1:2, 1977, 3–49; Afaf Lutfi Marsot, *Egypt in the Reign of Muhammad Ali*, Cambridge: Cambridge University Press, 1984; Ali Barakat, *Tatawur Al-Milkiya Al-zira'iya fi Misr wa Atharahu 'Ala Al-haraka Al-siyasiya, 1813–1914 [The Developments of Agricultural Landownership in Egypt and Their Impact on the Political Movement 1813–1914]*, Cairo: DWQ, [1977] 2017; Zain Al-Abideen S. Nigm, *Al-Dawla wal Mujtama' Fi Misr Fi Al-Qarn Al-Tasi' 'Ashar [State and Society in Egypt in the Nineteenth Century]*, Cairo: DWQ, 2007; Kenneth Cuno, *The Pasha's Peasants: Land, Society, and Economy in Lower Egypt (1740–1858)*, Cambridge: Cambridge University Press, 1993.

into the 1840s. Departing from the overarching question of how fossil-fuelled technologies made their way into Egypt on their way to global dominance, the two realms of landholding and regional wars, even if earlier historicised separately, here present themselves as one, combining global geopolitical with socioecological aspects of this history.

I start with a brief outline of the Syrian wars and the British-led European reactions culminating in the free-trade treaty of Balta Liman in 1838. In the next section, I flesh out my argument for rereading the Pasha's wars in Syria as mainly driven by his search for coal mines and, potentially, a strategic vision for an expanded Pashalik of Egypt and Syria with superior industrial potential, where Syria was to provide local cheap coal. Later sections are dedicated to explaining how the socioecological process, in the early 1840s, of the devastating cattle epizootic in Egypt – which progressed and dialectically coevolved with this redistribution – can explain further industrial and agrarian production developments associated with coal-based technologies and/or their absences. The complex chain of causality allows for a nuanced reading of Mehmed Ali's last decade in power – what I call the power crisis years – and sets the stage for understanding how steam power eventually made its way to Egypt, steadily and systematically, across later decades.

Once Mehmed Ali established a considerable grip over Egypt, with the massacre of the Mamluks in 1811, he set out on several successful regional military campaigns, becoming by the mid-1830s the governor of a mini-empire within the Ottoman realm. In the literature, the motives of this expansion are many, and arguments for and against a purposeful expansionist programme that sought independence from the Ottoman Empire abound.[2] What concerns us is that by 1835, the Pasha had overtaken large swathes of mostly Arabic-speaking regions in the Ottoman Empire that were also some of the wealthiest provinces: the Hijaz, Sudan, most of Yemen and eastern Arabia, Crete, and Syria. Some years later, his armies advanced to Anatolia, threatening the capital itself.[3] The two Syrian wars and the occupation of the Syrian

2 Some of these motives will be discussed shortly. A most thorough and balanced discussion on this question is in Khaled Fahmy, *All the Pasha's Men: Mehmed Ali, His Army and the Making of Modern Egypt*, Cairo: AUC Press, 2002, Chapter 1. See also: Marsot, *Egypt*, Chapter 9.

3 Fahmy, *All the Pasha's*, 38.

provinces, which lasted nearly ten years, included four spectacular victories of Ali's armies over the Ottoman imperial armies.

In November 1831, the cities of Jaffa, Haifa, Sidon, Tyre, Beirut, Tripoli, Latakia, Jerusalem, and Nablus all fell to Ibrahim Pasha, Ali's older son. After six months of besieging the strategic port city of Acre, it fell in May 1832 to Ibrahim. In June and July 1832, his army took Damascus and reached the northernmost borders of Syria, storming in with two armies, including an Ottoman army that the Sultan dispatched to halt Egyptian advances. Mehmed Rashid Pasha, the Grand Vizier, led a third and greater force, which was defeated on 21 December 1832, and he was captured during this battle, which took place in Konia, in the heart of Anatolia.[4] The Russians rushed to Istanbul the same month, offering defensive alliance with the Ottomans against Mehmed Ali. The Peace of Kutahia was signed in May 1833 with European support, mainly Russian and French. Short of a peace treaty, it was an agreement between the Sultan and his rebellious Wali to end the war in Syria, after giving him official recognition of his hold over Egypt, Crete, and Hijaz, and giving Ibrahim control and tax collection rights over parts of the Levant.[5]

Meanwhile, Lord Palmerston, the British foreign minister (in office 1827–9; 1830–4; 1835–41) felt the growing problem of Ali's expansion that threatened the integrity of the Ottoman Empire and brought a looming Russian influence to the region, in its turn threatening Britain's route to India.[6] He went as far as accusing the Pasha of secretly having 'the best understanding with Russia', conspiring together against British interests.[7] Palmerston had later, in 1835, refused Ali's approaches for gaining British support for independence from the Sublime Porte. When, in May 1838, Mehmed Ali voiced to all parties

4 Ibid., 61–73.

5 Marsot, *Egypt*, 227–30; Fahmy, *All the Pasha's*, 66–9.

6 A major theme in British foreign policy towards the Ottoman Empire over the first half-century was to keep French and Russian influence in Constantinople checked, in order to avoid their use of Ottoman support in any potential military moves against India. See Jonathan Parry, *The Promised Lands: The British and the Ottoman Middle East*, Princeton and Oxford: Princeton University Press, 2022. On the British foreign policy's overestimation of Russia's influence in the Levant, in the 1830s, see John Darwin, *The Empire Project: The Rise and Fall of the British World-System, 1830–1970*, Cambridge: Cambridge University Press, 2009, 29–31.

7 Marsot, *Egypt*, 236.

his insistence on independence, offering 3 million pounds to the Porte as a price, Palmerston's response was to offer British military support against 'the Dismemberment of the Turkish empire'.[8] This dismemberment of the Ottoman Empire, Jonathan Parry's recent extensive investigation of British diplomatic archives reveals, included fears over Mehmed Ali's potential expansion into Iraq and Arabia, as far as Bahrain, which the British navy in India decided in 1838 to defend in case of an Egyptian attack. The next year, a British position took form, appealing to the major powers of Europe to prevent Mehmed Ali from taking the Arabian Gulf.[9]

Exploiting the Ottoman fears of the Pasha's persistence on independence, and farther expansion into the Ottoman heartland and potentially across the Arabian Peninsula and the Gulf, the British negotiated the free-trade Treaty of Balta Liman in the summer of 1838 with the Sultan. The treaty, said to have 'put the nail to the coffin of all monopolies within the Ottoman Empire', was presented to the Turks as the strongest possible move against Ali, to undermine his economic capacity for raising and maintaining armies.[10] In Lord Palmerston's words, even if the treaty was, he claimed, going to increase Egypt's and Ottoman revenues over time, 'it may for the moment paralyse Muhammad Ali's scheme of finance'.[11] Marsot estimates that the Ottomans agreed to the free-trade treaty that dismantled their own monopoly systems across the empire, under Ali's existential threats, hoping that it would harm the latter even more, especially if he defied it – in such a case, putting him in conflict with the European powers. Mehmed Ali indeed ignored the treaty. He set off on a collision course with the Sultan, culminating in the battle of Nezib on the Turkish borders in June 1839. After a decisive victory, he restrained his son Ibrahim from storming Konia again, fearing the reaction of the European powers if he proceeded farther into Turkish territories. With this political manoeuvre, he was hoping to gain recognition of his independence from the powers. England, however, in response to his new demands, fell back to negotiations between all parties involved, and, by 15 July 1840, the London conference resulted in the 'Pacification

8 Ibid., 238.
9 Parry, *The Promised*, 136–43.
10 Marsot, *Egypt*, 238.
11 Cited in ibid., 239.

of the Levant' agreement, which Mehmed Ali defied again, refusing rights over Acre and southern parts of Syria – and Egypt for him and his house – in return for withdrawing from Greater Syria. Soon after, a British fleet equipped with steamers was dispatched, and Beirut was bombarded on 11 September 1840. An Anglo-Turkish force landed, and revolts erupted across Syria, under the influence of years of over-taxation and other coercive practices by the Pasha's occupation going into its tenth year, combined with Anglo-Ottoman agitation against the Egyptians that included providing the rebels with guns. Syrian provinces quickly fell one after another.

It was British steam power and coal in the form of superior manoeuvring capacity of war steamers – independent of the need to use wind to turn and move the ships – that brought decisive victory to the British fleet. The leading vessel in the fleet, the *Gorgon*, was dubbed 'the first true fighting steamship' in history.[12] Beirut was first bombarded in early September 1840, and then, as the squadron of steamers pressed on southward, other coastal fortified cities fell to the superior naval fighting technology. The fall of fortified Acre dealt the most devastating blow to Ibrahim's army, in November 1840. The *Gorgon* and the three other steamships in the naval force led the attack from the sea on Acre and manoeuvred easily away from the shots coming from the defenders, while the fleet bombarded them. Suddenly, one shot from the British side found its way to a central gunpowder magazine. Two regiments from Ibrahim's army were annihilated in seconds, 'and every living creature within the area of 60,000 square yards ceased to exist', and between 1,200 and 2,000 persons were dead.[13] A few weeks later, the British fleet appeared outside Mehmed Ali's palace windows in Alexandria.[14]

The imperial dream was over, and the free-trade treaty of Balta Liman had to be fulfilled. How did the question of motive power – beside its prominent appearance on the battlefield – feature in the Syria affair? I turn to this question now before moving back to Egypt, and how the

12 David K. Brown, *Before the Ironclad: Development of Ship Design, Propulsion and Armament in the Royal Navy, 1815–60*, London: Conway Maritime Press, 1990, 61.

13 On the steamships in Britain's war on Ibrahim Pasha's army (including primary material quoted, and a fuller description of the battle and its immediate aftermath), see Andreas Malm and the Zetkin Collective, *White Skin Black Fuel: On the Dangers of Fossil Fascism*, London: Verso, 2021, 344–6.

14 Marsot, *Egypt*, 244–6.

wars that sought expansion to feed accelerated commodity production with coal landed the Pasha in the 1840s power crisis.

In Syria's Coal Mines, 'preoccupied with his industrial thoughts'

The leakage of manpower to the Levant in the late 1820s and the coal mines of Mount Lebanon stand as two strong potential motives for the Syrian campaigns of the 1830s. Khaled Fahmy argues convincingly that the Pasha had, for decades, before moving the troops in 1831, eyed Syria for several reasons, including manpower and superior timber, and that the question of peasants fleeing to Syria was merely a pretext for the campaign in 1831. He further notes that coal – particularly its usage for gunpowder – was another reason, though an unlikely one, for staging a costly military occupation. Fahmy argues, instead, with a wealth of archival material, that the decision was taken in 1831 because the Pasha was acting pre-emptively upon the mistrust he felt the Sultan had gathered against him by then, with Syria being the place from which the Sultan's attack was anticipated to come.[15] While I agree with this potential geostrategic trigger for the campaign, I believe that the peasants who fled to Syria, and the coal mines of Mount Lebanon even more so, are compelling yet underestimated motives.

Concerning the peasants who fled to Syria, we know that *tasahub* (peasants leaving their homes and villages to flee taxes; more about them in the next section) extended outside Egypt, as far as Syria and other countries and provinces.[16] The 6,000 fleeing Egyptian peasants who took refuge in Syria at the turn of the 1830s were the pretext that Mehmed Ali declared for attacking Acre in 1831, claiming that its

15 Fahmy, *All the Pasha's*, 40–55.

16 About peasant flight elsewhere, Thomas Fowler, in 1861, stated as common knowledge, beside the peasants who fled to Syria from the Delta in the late 1820s, that some peasants from Upper Egypt fled to Abyssinia 'and neighbouring countries', as a cause of more land lying uncultivated in the late 1830s: Thomas Fowler, *Report on the Cultivation of Cotton in Egypt*, Manchester: J. J. Sales Printers, 1861, 7, 9. For more on peasants' flight, *tasahub*, including further relevant literature in Arabic and some primary material, see Marwa Rifaat Saleh, *Jara'im al-Ta'addi 'ala al-Aradi al-Zira'iyya wa Hadr Mawarid Al-Miyah [Crimes Against Agricultural Land and Water Resources, 1805–1914]*, Cairo: GEBO, 2021, 105–9.

governor offered them protection.[17] Peasants from Sharqiya, the eastern Delta, in particular, who fled to Al-Sham (Syria, Lebanon, Palestine) started showing up as a crisis in the Pasha's orders in March 1829. Their clandestine emigration and actions taken for remedying the situation escalated quickly. That month Mehmed Ali ordered Ibrahim to deploy 150 soldiers on travel routes east of Cairo to intercept fleeing 'Sharqiya peasants'.[18] In late July 1829, he ordered the return of travellers on the Suez and Syria routes, who possessed passports for Hajj, 'because such passports do not prescribe these roads. It is not the Hajj season', and 'do not allow folks to leave Egypt, they are categorically prohibited'.[19] In December 1829, an official from Syria informed the Pasha that half the peasants of Sharqiya who fled east returned to Egypt and requested camels to bring the other half back. Mehmed Ali called for a meeting with his aides 'on how best to return the Sharqiya folks from Al-Sham, and to brief him on the means needed'.[20] In mid-March 1830, he allowed 'rich farmers' in Sharqiya to build *'ezbas* and other hamlets for labourers who walked long distances to work the land, in a move possibly tied to enhancing conditions for Sharqiya's potential escapees.[21] Six days later, Mehmed Ali sent Acre's governor a letter accusing his *sheikh al-balads* (local rural administrators) of holding back returnees who originally fled from Sharqiya. If this blocking did not stop, he said, he was going to 'send special personnel to bring them [Acre's *sheikh al-balads*] along with the Egyptian peasants returning, and discipline them'.[22] The Syria campaign started a few months later.

If this episode of fleeing Sharqiya peasants was merely a pretext for the attack, why would the Pasha waste ink on these internal measures

17 Claim on this being the pretext for the Syrian campaign, in Fahmy, *All the Pasha's*, 41.

18 Order to Ibrahim Pasha dated 7 Ramadan 1244 H [12 March 1829], in Rauf Abbas (Editor), *Al-Awamir wa Al-Mukatabat Al-Sadira Min 'Aziz Misr Mohamed Ali [Orders and Correspondences Issued by the Wali of Egypt Mohamed Ali] Vol. I*, Cairo: Dar Al-Kutub, 2010, 253, item 1197.

19 Order to Mohammed Agha dated 21 Muharram 1245 H [22 July 1829], in ibid., 269, item 1277.

20 Order to *Ma'moor* of half Sharqiya dated 22 Jumada Thani 1245 H [18 December 1829], in ibid., 282, item 1338.

21 Order to Sharqiya police director, 21 Ramadan 1245 H [15 March 1830], in ibid., 289, item 1379.

22 Order to Acre governor, 27 Ramadan 1245 H [21 March 1830], in ibid., 290, item 1384.

for addressing the crisis? Furthermore, if we factor in the timing of the harsh measures the Pasha imposed on internal migration, in 1829 (a village passport system for all persons; more about this in the next section), it will seem likely that he genuinely feared a broader mass fleeing of peasants from the Delta to the Levant. We can safely assume that, in principle, governors of the Levant were welcoming of such migrants, who were to add to their own rent and manpower base. Perhaps a certain osmotic pull was building up between the eastern regions of the Nile Delta and Greater Syria, escalating quickly, and needed to be addressed.

Concerning the coal mines that Al-Sham offered, at least as early as late 1823, at a time when the Pasha was avidly dispatching European coal and minerals prospectors across Upper Egypt, he knew about 'coal discovered in Mount Lebanon'. He wrote to the governor of Sidon that year, thanking him in advance for 'appointing a chemist to extract this coal and for [intending to] send enough quantities to Egypt every year'.[23] The extant orders from the Pasha are thin on the Syrian coal for the rest of the 1820s. Perhaps the governor of Sidon did not send any coal. Suddenly, with the start of the campaign in late 1831, the coal mines reappear, and the European experts the Pasha sent to Mount Lebanon from Cairo turn up, giving the impression that they marched with Ibrahim Pasha's army.

In the fall of 1831, James Brettell, from Cornwall, according to the American consul in Cairo, started exploring, boring, and mining in Mount Lebanon for the Pasha.[24] Then we hear more in a letter from early August 1835, from Mr Brettell in Mount Lebanon, to Omar Affandi, director of supplies in the Pasha's government in Cairo. He reported on his explorations of coal and iron. Many mines were found. The coal discovered, however, was heavy in sulphur, and, as such, not fit for the foundries, but might prove useful in steam engines. He promised to keep digging to find more coal. Mr Brettell wanted information on how much coal Cairo needed, and he was to endeavour to send the quantities specified, even if he was not convinced that the coal found was of tolerable quality. He was to enhance the road between the mountain and the coast as much as he could on a limited budget, for

23 Viceroyal letter to the Wali of Sida (Sidon) dated 22 Safar 1239 H [27 October 1823], in ibid., 101, item 486.

24 Richard C. Taylor, *Statistics of Coal*, Philadelphia: J. W. Moore, 1848, 633.

transferring coal using camels, mules, and donkeys. A certain Osman Qawass learned mining from an English miner, Richard Thornley, but Brettell needed 'more like him [i.e. Qawass] or four or five English miners' to supervise the men employed and help avoid accidents. In short, by 1835, Mr Brettell was busy digging and bringing coal – with questionable quality – and slowly expanding the infrastructure and specialised labour for the works, using a limited budget.[25] This does not mean, however, that coal was not a priority. Ibrahim Pasha decided in 1835 to oversee the coal-mining process, refusing in correspondence with Cairo to leave this to Syrian princes whom he thought were 'keen on pillaging and embezzling official monies'.[26]

British estimates for the quality of Syrian/Lebanese coal were optimistic, at least until 1837. That year, based on Alexander Galloway's reporting, and that of others, to the House of Commons, it was thought in Britain that 'very superior coal, and very superior iron' were discovered in Syria, and thousands of tons of coal were transferred to Alexandria in the mid-1830s.[27] The House of Commons reached the conclusion that such coal was useful for steam navigation, but was 'in the progress of experiment' by the Egyptians.[28]

The American consul in Cairo, a few years later, gave a much bleaker picture on the quality of this coal. He said that, once it was shipped to Alexandria, engineers were ordered to try it in the few steam engines in the country. This coal was 'condemned and absolutely rejected by every engineer in Egypt'. It destroyed metal surfaces it burned within, 'and several boilers, grates, etc, were ruined'. It was found useful only for working lime, but then with lime cheaper fuels were available.[29] In September 1833, the Pasha ordered deeper digging at the coal pits in the region, 'for perhaps better quality is to be found [. . .] in such a case send

25 Amin Sami, *Taqwim Al-Nil 2 [Survey of the Nile Vol. 2]*, Cairo: Matba'at Dar Al Kutub, 1936, 442–8.

26 Abdullatif M. Al-Sabagh, 'Juhud Mohamed Ali Al-Ta'diniya bil-Sham Ibban Al-Hukm Al-Masri (1831–1840) [Mehmed Ali's Mining Efforts in the Levant During the Egyptian Rule (1831–1840)]', in *Islah Am Tahdith? Misr fi 'Asr Mohamed Ali [Reformation or Modernisation? Egypt in the Age of Mehmed Ali]*, edited by Rauf Abbas, Cairo: Supreme Cultural Council, 2000, 370.

27 House of Commons, *Report from the Select Committee on Steam Communication with India*, London: House of Commons, 1837, 66.

28 Ibid., 226.

29 Taylor, *Statistics of Coal*, 633.

forth a sample of that too'.[30] However, the whole enterprise was eventually abandoned, and engineer Brettell returned to Cairo in 1837, where he stayed until 1840, before apparently returning to England with the end of the Levant conquest. Probably before 1840, Mehmed Ali gave the American consul a box of the Mount Lebanon coal, asking for a second opinion from the United States. The latter forwarded the samples to a Professor Renwick in New York, and to several engineers at the West Point foundry. All the American experts reportedly declared it 'unfit for any purpose'.[31]

Regardless of the disappointing quality of the Mount Lebanon coal, the coal miners and transporters of coal from the mountain to the coast in Beirut were at the heart of the popular uprising of 1840 in Al-Sham that Britain and the Ottomans used and/or supported to end the Pasha's attempts at reconfiguring the Ottoman Empire.

Around May/June 1840, the mountaineers rallied an imposing force of 40,000 men. They reportedly defeated one of Ibrahim Pasha's regiments, besieged Beirut, and were joined by Bedouin groups.[32] The British navy joined force, leading to the start of Ibrahim's ultimate defeat and retreat to Egypt later that year.[33] This uprising and several others at the time were said by a contemporary commentator to have been 'in consequence of the mountaineers having been obliged to work in the coal mines'.[34] As they laid siege to Beirut, the Mount Lebanon rebels, in a petition addressed to local authorities working with Ibrahim Pasha, said that although they joined the troops that had earlier vanquished revolts in Saffert, Nassriya, and Nablus, they were later made to pay *firda*

30 Order to deputy of the Acre governorate, 3 Jumada Awal 1249 H [17 September 1833], in Abbas (Editor), *Al-Awamir Vol. I*, 374, item 1829.

31 Taylor, *Statistics of Coal*, 633. This bleak picture on the quality of the Mount Lebanon coal from the Americans goes against the findings of Al-Sabagh's rare study on Mehmed Ali's Syria coal mining, as he concluded that Egypt 'got most of her needs of coal' from such coal mines. See Al-Sabagh, 'Juhud', 373. However, it is very likely that this conclusion is wrong. The study relied *exclusively* on Egyptian coal-mining correspondences from the period accessed at DWQ and seems to have ignored evidence it brings on the many kinds of problems surrounding production. It remains, however, a rare source of relevant primary material.

32 'Spirit of Foreign News', *Cheltenham Examiner*, 5 August 1840.

33 Andrew A. Patton, *A History of the Egyptian Revolution: From Arab and European Memoirs, Oral Tradition, and Local Researches, Vol. II*, London: Trubner & Co., 1863, Chapters 12, 13.

34 Ibid., 179.

(individual taxes), were mistreated and had their women mistreated, and were 'hung upon trees'.[35] Their grievances, however, focused on the coal mines more than anything else. The letter signed 'by the inhabitants of the mountain in general' continued:

> When the coal-mine was discovered in the mountains, the mountaineers were ordered to work it and to furnish the necessary implements without being paid for them; and people were sent to inspect these works and workmen and others were paid to convey the coal to Beyrout; but this was of small amount, and we were obliged to provide for the rest of the payment at our cost, and to furnish posts and sacks for this mine, and we were only paid the quarter of the price of them, but the expenses of conveyance from the villages to the mine were not paid to us. If we were to enter into the details of all these vexations it would be too long, and we do not reckon the blows with sticks, and the dishonours which were done us, as if we have been Egyptian fellahs.[36]

This comparison they conjured to 'Egyptian fellahs' has more to it than the image it gives on how Mehmed Ali intended to co-opt the mountaineers into his notorious and exhaustive methods of appropriation of wealth endured by the Egyptians for decades. A contemporary commentator gave a nuanced version of a grander vision of an Egypt-Syria project that might have been in the mind of the Pasha since the first day he sent his troops east. Auguste Colin, a French lawyer and journalist, captured as a 'commercial absurdity' the unequal exchange at the time between Egypt and Europe, whereby Mehmed Ali sold cotton to Europe and then bought fabrics from it, while sound economic logic would have necessitated that he manufactured all the cotton into fabrics. He was lacking in the coal and iron necessary for the industry, however.[37] The solution he resorted to, Colin estimated in 1838, as if turning hypothetical thoughts in the Pasha's mind into words, was to transport

35 'Syria', *Morning Post*, 31 July 1840, translated letter from the Inhabitants of the Mountain to Emir Amin, dated 11 June 1840.

36 Ibid.

37 Auguste Colin, 'Lettres sur l'Égypte: Industrie Manufacturière', *Revue des Deux Mondes* 14, 1838, 517–31, 530–1.

all his industries to Syria, where he 'hopes to find metals and coal'.[38] A never-realised industrial revolution fed by metals and coals extracted by more extremely exploited populations, which hinged on a frustrated attempt at building an Egypt-Syria kingdom, unfolds in Colin's words, which deserve to be quoted at length:

> Indeed, research has shown the presence of coal in the Taurus chain [. . .] The coal was found, by the probe, at a depth of forty feet, in a layer of bituminous shale. But unfortunately this coal is of a very inferior quality, and not very suitable for industrial use. Already, since 1833, when Mohammed Ali, victorious over the Sultan, claimed, in diplomatic negotiations, the cession of the province of Adana, rich in wood and mines, he was preoccupied with his industrial thoughts. Master of this province, he made mineralogical researches there. It was found that there were eight mines already exploited and providing different metals [. . .] Until now, what has hindered and delayed Mohammed Ali's projects is the constant state of unrest on the Syrian border and the periodic revolts of the inhabitants. *That is why he is in a hurry to finish it, even by war. If more stability allowed him to carry out his plans, Egypt would then return to its agricultural specialty; manufacturing industry would be established in Syria; the two regions would complement each other industrially, one more specially devoted to agricultural and chemical industry, the other more specially devoted to manufacturing and mechanical industry.* This is how what we have said about the industrial destiny of Egypt would be realised.[39]

This is likely to have been the case: to initiate wide-scale industrialisation in Greater Syria, a land that offered a promise of abundance in coal and minerals, with a population that could be coerced into mining the coal and iron needed for industrialisation. The undoing of this hypothetical vision, however, seems eerily similar to how the industrial project was undermined in Egypt with the power crisis of the early 1840s. We turn to this next.

38 Ibid.

39 Ibid., 531. Emphasis added.

The Squeeze, and the Fightback

If British-Ottoman coercion, in combination with local resistance to the occupation, including the uprising of the coal miners of Mount Lebanon, seems to have brought Mehmed Ali's Syrian industrialisation project to an end before it began, its undermining might have owed no less to longer-term structural factors related to how Ali powered his military machine from Egypt. The contradictions of the Wali's mode of labour exploitation had by the 1830s left the Egyptians thoroughly depleted.

As early as 1811, Mehmed Ali established full state control over cash crops.[40] After defeating the Mamluks, and in the middle of the Napoleonic wars that increased international demand for grains, he banned all trade in cereals in Upper Egypt, becoming the sole buyer and reseller of the crop.[41] Other crops quickly followed: the Delta's rice in 1812, sugar in Upper Egypt in 1815, and sesame, indigo, cotton, hemp, wheat, barley, and beans, among other crops, by 1816.[42] This monopoly system was accompanied by meticulous oversight and control over the rural subjects and their land. Peasants were told what exactly to grow and when. They were issued domestic passports for movement and relocation outside their villages, and were instructed to get married and reproduce at an early age to bring more working hands to the land (more on this shortly). Beside the monopoly itself being an expensive export tax of sorts, all kinds of demands and taxes were unleashed on the population.[43] By cutting out middlemen (mainly cutting the throats of

40 According to many sources (Laura Panza and Jeffrey G. Williamson, 'Did Muhammad Ali Foster Industrialization in Early Nineteenth-Century Egypt?', *The Economic History Review* 68:1, 2015, 79–100; Owen, *Cotton*; Marsot, *Egypt*). Kenneth Cuno, however, estimates that monopoly originated partially as early as 1808. See Cuno, *The Pashas'*, 104, 160–1.

41 Prices of cereals on European markets, under the impact of the war in Europe, more than quadrupled in 1810–13. This cereals trade boom carried on in Egypt until 1820. See Fred H. Lawson, 'Rural Revolt and Provincial Society in Egypt, 1820–1824', *International Journal for Middle East Studies* 13:2, 1981, 131–53, 143.

42 Roger Owen, *The Middle East in the World Economy*, London: I. B. Tauris, 2002, 66; Marsot, *Egypt*, 153. Ahmed Al-Hitta proposes slightly different dates for when monopolies in different commodities started, but all still within the late 1800s and early 1810s: Ahmed Al-Hitta, *Tarikh Misr Al-Iqtisadi fi Al-Qarn Al-Tasi'a 'Ashar [Economic History of Egypt in the Nineteenth Century]*, Cairo: Dar Al-Kutub, [1958] 2012, 146–7.

43 Panza and Williamson, 'Did Muhammad', 82.

Mamluks in 1811) who were privileged with the *multazim* (tax collector's) office, all lands fell to direct state control – Upper Egypt in 1812 and the Delta in 1814 – as such making it possible to increase direct land tax revenues, amounting to at least half of total state receipts.[44] Expansion in agricultural production to increase returns from cash crops required the shift to perennial irrigation for the first time in Egypt – where, traditionally, irrigation relied on the rhythm of the annual Nile inundation – and the digging of new deeper summer canals, using corvée labour on unprecedented scales.[45] The building of armies for military campaigns and conquests to provide inputs for and protect this accelerated commodity production required the forced recruitment of soldiers from among the Egyptian peasantry. The labour for the new manufactories that sprang up across Ali's reign, estimated between 180,000 and 230,000 workers – roughly 20 to 30 per cent of men over fifteen years of age – was coercive across the board, even if paid.[46] For irrigation-related corvée alone, according to some estimates, between 1820 and 1831, around 67,000 men were needed annually for canal digging, and 400,000 other men for seasonal canal clearing, among a population that could have reached 5 million by 1830.[47] At least 100,000 peasants fuelled the Pasha's armies in the 1830s, including the troops that fought the Ottoman armies in northern Syria.

The literature converges on two main strategies for popular resistance against this brutal process: outright revolt and *tasahub* (peasant flight), beside the occasional peasant sabotage of government irrigation infrastructural works and crops.[48] At least five uprisings erupted between 1820

44 Owen, *The Middle East*, 65.

45 Alan Mikhail, 'Unleashing the Beast: Animals, Energy, and the Economy of Labor in Ottoman Egypt', *American Historical Review* 118:2, 2013, 317–48.

46 See Marsot, *Egypt*, Chapter 8, on industrial labour population estimates.

47 Cited in Cuno, *The Pasha's*, 122. Cuno (with Reimer) also defended the 5 million census estimates as accurate; see Kenneth Cuno and Michael Reimer, 'The Census Registers of Nineteenth-Century Egypt: A New Source for Social Historians', *British Journal of Middle Eastern Studies* 24:2, 1997, 193–216.

48 Lightly hinted at in Cuno, *The Pasha's*, 122–3: 'Resistance also seems to have taken the form of sabotage. After the ruinous flood of 1829, for example, it was rumoured that "many cultivators" secretly cut the dikes that protected the fields, adding to the destruction'. The many orders the Pasha issued on punishing peasants who uprooted cotton shrubs in their lands across the years and left their cotton lands fallow, embezzling from the land, or replaced it with other food crops, also hold witness to this other more subtle form of sabotage. See for example: Abbas (Editor), *Al-Awamir Vol. II*, items 2613, 2797, 2823, 2834, 3377.

and 1826, the largest being in Upper Egypt. At least one was witnessed across ten villages in Sharqiya in August 1826. Another uprising erupted in Munufiya, middle Delta, in 1823. Almost all were attributed to factors including coercive labour, conscription practices, and taxation.[49] Uprisings in the early 1820s were accompanied with, then dwarfed by, reports on peasant flights, across the 1820s and 1830s, suggesting voluntary displacement as the more sustained and successful peasant resistance and survival strategy. Kenneth Cuno's extensive study of the Egyptian peasantry under Mehmed Ali includes overwhelming evidence on progressive impoverishment and peasant flight from villages since the early 1820s, escaping forced conscription, corvée, and taxes in arrears.[50]

Mehmed Ali's reign was frequently dotted with stories of deserted villages and fleeing peasants who left their fields untilled, according to Roger Owen, who estimates that by 1837, rural conditions were much worse than they had been over a very long time, in association with increasing rates of flight. This misery, he contends, was largely the result of the agrarian-produce monopoly system.[51] By 1832, the taxes in arrears across Egypt, in no trivial part due to *tasahub* and military and labour demands on the peasants, amounted to 75 per cent of a full year's taxes. The peasants' way of pushing back was to flee to cities and other rural centres, as gleaned from several governmental orders to stop this clandestine form of internal emigration.[52] By 1838, there were reports of many villages in the Delta falling as much as four years behind in tax collection.[53] This trend is supported by rough estimates of Mehmed Ali's overall annual revenues volume that reached its peak in 1829, then took a downturn across the 1830s and 1840s.[54]

49 Cuno, *The Pasha's*, 125, 138. See also on peasants' revolts sources listed in Cuno, *The Pasha's*, 241, endnote 15. On the social groups in Upper Egypt's revolts in the early 1820s, Lawson argues against the mainstream assumption that they were peasant revolts – so described by 'Rivlin, Baer, Richards, and virtually every other student of Egyptian history' – and instead places these revolts (centred around Qina in 1820–4) in the hands of discontented village artisans and pieceworkers who manufactured and processed agrarian products for local consumption (drowned in increasing imports from Europe) and exports. Lawson, 'Rural Revolt', 145.

50 Cuno, *The Pasha's*, 117, and Chapter 8.

51 Owen, *Cotton*, 58.

52 Government reports on *tasahub* cited from: Helen Rivlin, *The Agricultural Policy of Muhammed Ali in Egypt*, Cambridge, MA: Harvard University Press, 1961, 131.

53 Owen, *Cotton*, 59.

54 Cuno, *The Pasha's*, 118, Table 6.3.

It was also in 1829 that several procedures were attempted to address this leakage from the Pasha's exploitative mode of agrarian commodity production. Beside the earlier widespread governmental practices of tracking down and returning fleeing peasants to their villages, Mehmed Ali promulgated in 1829 a system of domestic passports (*tazkarah*), which needed to be issued for any persons who wanted to leave their village for any reason. In the document, *sheikh al-balads* wrote the peasant's name, skin colour, age, and bodily description, along with the reason for leaving the village and for how long. Local administrators were instructed to inspect the passports of any strangers in their districts. Six months' to two years' imprisonment was levied against any person in possession of a fake passport.[55] The stated purpose in the first decree regulating this process was to 'remove the word escapee [*harib* in Arabic] entirely from existence, as he [the Pasha] wishes'.[56]

The same year, Mehmed Ali issued several orders addressed to village sheikhs: every person under their oversight should get married as soon as they reach puberty. If the youth could not afford marriage, the village sheikh was to provide help.[57] This support was financial, but coercion also might have been used. A visitor to Egypt in 1831 went so far as to suggest that a village sheikh gathered all the boys in his district and told them that they must all get married before turning fourteen, or else they would face beating and other severe consequences.[58] Orders were issued preventing abortion, decreeing a prison sentence of two years for violators. Other orders appear on the record, between 1829 and 1835, about sentences and measures to be taken to deal with the problem of peasant self-maiming that was induced to avoid corvée and military conscription, rendering the rural subject in the process a less efficient hand for tax and crop-production purposes. The Pasha outlawed the selling of rat poison, which peasants used to induce blindness.[59] Some of the self-maimed were arrested and sentenced to death, others were imprisoned

55 Cuno, *The Pasha's*, 124; Fahmy, *All the Pasha's*, 106.

56 Order to Katakhda Bey dated 11 Shawal 1244 H [15 April 1829], in Abbas (Editor), *Al-Awamir Vol. I*, 258, item 1220.

57 Al-Hitta, *Tarikh*, 122, citing *Al-Waqa'i' Al-Masriya*, 5 Rajab 1247 H [9 December 1831].

58 P. H. Hamont, *L'Egypte sous Mehemet Ali*, Paris: Leautey et Lecointe, 1845, 311.

59 Order to Katakhda Bey in Cairo dated 17 Sha'ban 1245 H [10 February 1830], in Abbas (Editor), *Al-Awamir Vol. I*, 286, item 1363.

for life, and some were shipped to the factories and other production sites. Persons who became professional maiming practitioners were also sentenced to death, in occasionally grotesque procedures.[60] Mothers who blinded their young children were punished by drowning in irrigation canals.[61] Quarantine and vaccination systems – which, Cuno contends, were motivated by the need for labourers and taxpayers – were devised in the early 1830s, leading, for example, to the disappearance of smallpox by the 1840s.[62]

But, with the population and tax leakages growing with time, undefeated by these measures and others, as the Pasha's finances showed in the late 1830s – combined with the free-trade agreement of Balta Liman in 1838 – something else had to be devised. It is hard to decide which set of factors pushed Mehmed Ali to think of redistributing land: internal tax resistance leading to revenue and labour power leakage, culminating in spectacular losses in the form of uncollected taxes by the late 1830s, or the anticipated shifting regional and international terms of trade that were eventually going to be consummated, even if the Pasha resisted the treaty at first. A mix of the two categories of motives is likely the more probable scenario, even if Mehmed Ali had taken action towards land redistribution before Balta Liman.

In January and February 1838, the Pasha called a Council of State that the provincial governors attended. They were asked to devise solutions to remedy the crisis of uncollected taxes. Their suggested solution, from the toolbox of the Pasha's land policies, was to conduct a new cadastral survey, to be followed by a redistribution of tax-insolvent village lands, to villages that could shoulder that burden.

60 Order to the governor of Fayoum, dated 23 February 1830: Makkyiah, the wife of a hairdresser who practised blinding by rat poison to help boys evade conscription, was sentenced to crucifixion, 'at the market of Fayoum, and leave her corpse there for three days'. The boys who were blinded 'willingly' were to be imprisoned for life at Alexandria's prison. Abbas (editor), *Al-Awamir, Vol. I*, 288, item 1370. An apothecary was sentenced likewise to crucifixion 'for blinding babies by rat poison'. Governors across the country were ordered to levy the same sentence in comparable crimes. Ibid., 374, item 1831.

61 Decrees and judicial verdicts issued 1829–35 are cited in Al-Hitta, *Tarikh*, 122.

62 Cuno, *The Pasha's*, 124. It can be hard though to pin down mounting quarantine efforts as a purely domestic policy. For international trade as a motive for the 1840s Egyptian quarantine policies and efforts, see Vladimir Hamed-Troyansky, 'Ottoman and Egyptian Quarantines and European Debates on Plague in the 1830s–1840s', *Past and Present* 253, 2021, 235–70.

But the Pasha apparently needed a quicker fix.[63] Instead, as the British consul general reported to Lord Palmerston in February 1838, he reassigned whole villages to new proprietors who became tasked with squeezing out uncollected taxes for the state.[64] Bowring described the new proprietors as 'capitalists'. He had visited such redistributed estates, ranging between 300 and 800 acres, and thought the new enterprise was profitable to the estate holders and the Pasha.[65] The legal terms that emerged from these redistributed estates were the previously infrequently used terms *'uhda* (custody lands) and *muta'ahidun* (custodians).[66] Although land records show the occasional use of these terms earlier in the 1830s,[67] they saw a radical leap in use by early 1838. By 1840, the *'uhda* estates appeared in official records as a state-wide phenomenon. On 23 March 1840, the Pasha decreed that all *muta'ahidun* who grew rich during the war years should hasten to pay the accumulated taxes in arrears of their newly given estates and guarantee their future tax liability.[68]

In some ways, this was a return to the centuries-old *iltizam* (tax-farming) system that Mehmed Ali departed from early in the second decade of the century. The main differences between the pre–Mehmed Ali tax collectors' scheme and this renewed relegation of the tax collection system was that in the latter, the guarantors were not allowed the right to collect *fa'iz* (tax surplus); they had to dispatch all collected taxes to the government. For their trouble, they were allowed to set aside parts of the estates for their own cultivation (*usiya*), with usufruct rights, that amounted to 50 per cent of the estate in some cases, accompanied with the right to set aside some of the hands on their *'uhdas* for dedicated work on the *usiya*. Those labourers were secured against corvée and

63 Rivlin's position, which Owen agrees with: Rivlin, *The Agricultural*, 132; Owen, *Cotton*, 60.

64 FO 142/9, Campbell, 26 February 1838, cited in Owen, *Cotton*, 60.

65 Bowring, cited in Cuno, *The Pasha's*, 158.

66 Ghislaine Alleaume, 'An Industrial Revolution in Agriculture? Some Observations on the Evolution of Rural Egypt in the Nineteenth Century', in *Agriculture in Egypt: From Pharaonic to Modern Times*, edited by Alan Bowman and Eugene Rogan, Proceedings of the British Academy, Vol. 96, 1999.

67 The records (governmental decrees) uncovered by Al-Hitta hold evidence on whole villages given as *'uhda* to senior officials, European residents, and rural notables between 1831 and 1836. See Al-Hitta, *Tarikh*, 70.

68 Decree cited in Gabriel Baer, *A History of Landownership in Modern Egypt, 1800–1950*, Oxford: Oxford University Press, 1962, 13.

military conscription.[69] Administrative and law-enforcement powers were privatised, relegated to the new proprietors; they were to provide the peasants with working capital to cultivate crops – which were still dictated by orders from the central government – and to practise some judicial and executive powers over their subjects, with the government being a supreme arbiter over the two parties.[70]

Although conceived as a temporary arrangement to fix the crisis of taxes in arrears and alleviate the administrative costs associated with directly managing tax collection and providing agricultural working capital, by 1846, around 677,000 acres of Delta land were redistributed as *'uhda*.[71] As Owen hints, the crisis-born *'uhda* fix meant a partial displacement of the tax crisis to the upper class in Egypt.

Seen in combination with the other main mechanism of land redistribution during the early 1840s, the *chiflik* system, it becomes clearer how the Pasha's land redistribution plan worked for a while to preserve some of the benefits of the monopoly against the tide of forced British-Ottoman free trade.

Unlike the *'uhda* lands, which were generally within the tax-insolvent category, and some of which Mehmed Ali and his family took themselves, the *chiflik* estates were in all likelihood profitable fertile lands that grew cotton and other cash crops, including rice. Between 1838 and 1846, over 334,000 acres were redistributed as *chifliks*, of which Mehmed Ali himself took two thirds. The legal documents of the redistribution, thoroughly examined by Ali Barakat, show traces of the forced seizure of productive peasant lands as *chifliks*. Barakat, moreover, contends that, in all likelihood, the dispossessed peasants were forced to stay on the land as tenants and sharecroppers.[72] Devised and implemented in

69 Alleaume, 'An Industrial', 334.

70 Owen, *Cotton*, 60. With archival evidence, Ali Barakat maintained that judicial rulings by *muta'ahidun* were exclusively 'verdicts of first instance'. It follows that plaintiffs might have been able to go to higher-level courts outside the estate. Barakat, *Tatawur*.

71 Cuno, *The Pasha's*, 159.

72 There is supporting evidence for Barakat's hypothesis on forcing land-dispossessed peasants to stay on *chifliks* as tenants and sharecroppers. In a report from the American consul in Cairo at the turn of the 1840s, it is said that after a village is declared as part of a *chiflik*: 'soldiers are sent, who form a cordon around the doomed spot, and then commences the seizure of everything belonging to the villagers; their houses, lands, crops, trees, implements, cattle, poultry, seeds, stocks, clothes, food, and

tandem with the Balta Liman treaty of 1838, this expansion in the *chifliks* scheme was used as a defensive measure against the impending abolition of the monopoly system. By the mid-1840s, all 'peasant lands' that remained from the 1810s–30s Egyptian land system were only 47 per cent of all cultivated land in Egypt, while *'uhda* and *chiflik* lands together amounted to 49 per cent (36.82 *'uhda*; 12.1 *chiflik*).[73]

It should be pointed out that the redistribution of lands to the Viceroyal family was not a novel affair. In 1808, even before the full abolition of *iltizam* with the Mamluks massacre of 1811, Mehmed Ali wrote to his land tax officials asking for a special 'war/ammunition' (*zakhira*) tax. When they responded that many villages were in ruin, he demanded two lists, of productive and ruined villages. The 160 villages in the latter list he ordered to be reassigned to his 'family and court', that is, an early application of the likes of *chifliks* and *'uhda* as emergency measures for keeping the Pasha's grip on power. In the same year the *multazims* of the Behira province declined to send taxes to Cairo, so the Pasha declared their lands reassigned to his 'trusted circle'.[74]

In times of crisis, the redistribution of land was thoroughly deployed, and the free-trade crisis of the 1830s–40s was perhaps the greatest crisis the Pasha had faced since his early years in power. Mehmed Ali used land redistribution, and loopholes in the free-trade agreement of 1838, to evade the abolition of the monopoly system for a while. The treaty explicitly prohibited all monopolies within the Ottoman Empire. It abolished taxes imposed on British traders who wanted to buy commodities everywhere in the empire, and imposed instead the much lower duties of 3 per cent on imports (to Ottoman lands; all other internal imports taxes were abolished), 12 per cent on exports, and 3 per cent on transit.[75] When this

---

property of every kind, down to the distaffs and the cooking pots, which (saving what may have been plundered by the soldiers), are all confiscated to Mohammed Ali. This done, the villagers are compelled to till for the Pasha's profit that soil which was (and which, notwithstanding the imposture, undoubtedly still is) theirs, receiving wages at the extreme rate of 40 paras, (equal to 2.5d.) per diem for each able-bodied man, and less in proportion for the labour of women and children.' See George R. Gliddon, *No. 1: A Memoir on the Cotton of Egypt*, London: James Madden & Co., 1841, 25.

73 Cuno, *The Pasha's*, 164, Table 8.3, calculated from Rivlin and Barakat's extensive land-redistribution study.

74 Al-Hitta, *Tarikh*, 55–6.

75 Marsot, *Egypt*, 239. Panza and Williamson, 'Did Muhammad', 83.

was first conceived, the French ambassador at Istanbul described the treaty as 'a complete commercial revolution', with the taxes on some items it covered reduced tenfold. A decade later, Lord Palmerston boasted in the House of Commons that 'it is quite true that with no other country is our trade so liberally permitted and carried on as with Turkey'.[76] But, in early 1841, when Sir Charles Napier, the commander of the British fleet that retook Syria from the Pasha, landed in Alexandria to lead the negotiations on the end of monopoly, Mehmed Ali tried to talk him into a gradual transition from monopoly to free trade:

> *You had once in your country a monopoly of the East India trade. It required a great many years for you to do away with it, and you must allow me some time for the accomplishment of what I conceive to be beneficial.* When I can see my own interest, nothing will give me greater pleasure than to open the whole trade in Egypt, and on the Nile, to British commerce. What I expect is, that England will tell me what she wants, and I will do it. Our interests are completely identified. You want a passage to India. I want your trade, and just tell me what it is that you expect, and I shall be happy to enter into all your views, whatever they are.[77]

There were widespread European complaints about the Pasha's meddling in the liberal trade with Egypt after his meeting with Napier in early 1841, in which he requested more time before monopolies ended. In 1842, the French consul in Cairo reported his fears for how the Pasha remained – through indirectly controlling large expanses of land, and hence, commodities – able to evade the undermining of his monopoly system, at least in terms of Egyptian exports. Through 'distribution of villages by His Highness to the members of his family and the court', he remained in control of trade.[78] The small businesses that wanted to trade with Egypt, the consul said, were the most harmed, while large commercial houses preferred to trade directly with the government in all cases. Simultaneously, the 'provincial

76 Quoted in Marsot, *Egypt*, 239–40.

77 'Landing of Commodore Napier at Liverpool', *Leeds Intelligencer*, 24 April 1841. Emphasis added.

78 Cited in Marsot, *Egypt*, 251.

authorities' in Egypt kept hindering the free 'navigation and circulation of merchandise'.[79]

The Times, on 4 June 1841, reported on a rather sophisticated case of free-trade evasion by the Pasha. Linseed, the only available seasonal grain that spring, was sold by small farmers in baskets to individual European trade representatives in the Nile valley. The peasants received money perhaps for the first time in their lives, the reporter exclaimed.[80] But as the representatives of the British commercial house that bought the linseed from the villages were getting ready to leave Alexandria with their purchases, they wanted first to meet the Pasha, to deliver

> the address of the merchants of London expressing their most grateful acknowledgment for the protection he afforded persons carrying on commercial pursuits, and their fervent prayer that he may continue to develop the commercial energies of Egypt, unfolding to their fullest extent the blessings of prosperity and civilization, [but meanwhile] *he was issuing counter-orders to prevent the exportation of the linseed.* All the merchants concerned, and even the Consul-General of France, hastened to Bhogos Bey [a senior official] to request to be informed of this strange and contradictory veto at the 11th hour, and were told that it was the result of the Pasha's anxious foresight for the necessity of next year's seed-time etc [i.e. to use the linseeds for next year's cultivation].[81]

Besides this arbitrary and unique case of evasion by the Pasha, other traders who tried to buy commodities from villagers discovered that *muta'ahidun* (custodians of *'uhda* redistributed lands) were systematically prohibited from selling, and/or were told that the produce was taken by the government as taxes tied to their estates.[82] Even those who managed to make purchases found it hard to acquire river boats and crews to transfer their commodities to shipping in Alexandria. They

79 Ibid.

80 Cuno and Peter Gran, among others, rightly maintain that Egypt was before Mehmed Ali's monopolies an integral part of regional and international trade in commodities, in the eighteenth century and earlier. See Cuno, *The Pasha's*; Peter Gran, *Islamic Roots of Capitalism: Egypt, 1760–1840*, Austin: University of Texas Press, 1979.

81 No title, *The Times*, 4 June 1841. Emphasis added.

82 Owen, *Cotton*, 66.

were told, the British consul explained, that the Pasha prohibited the manning of any European traders' boats in direct contact with rural producers. Instead, they found it easier, considering transportation logistics, to buy directly from the Pasha. These cases, and many other examples of free-trade evasion, show how far Mehmed Ali went in using the redistributed lands and explicit intrigue to keep the monopoly system going.[83]

There were, however, other consequences for the changing patterns of landownership at the turn of the 1840s that concern us: those related to the earliest precursors of an animal/steam power transition in agriculture.

The Plague and the *Chiflik*

The redistribution of agricultural land across Egypt under the blows of the free-trade agreement and the undermining of the monopolies in the early 1840s formed, in the longer run, an entrenched landed nobility class. Over the next two decades, this class gradually became a pioneering force for introducing – along with European fossil and financial capital – steam technology to agriculture and related industries, for example cotton ginning and sugar refining. However, in the shorter term, the few years during which the land was redistributed, those noblemen were hard-pressed to substitute the leaked revenues of the Pasha from years of taxes in arrears over peasant land. The Wali was keen to see agrarian commodities flowing from the *'uhdas* and *chifliks* to his warehouses and silos to sell to British and other European traders, the latter now empowered more than ever with the new liberal terms of trade of Balta Liman, with a whole country's produce slipping gradually through his fingers.

What appears to be the disintegration of the Pasha's centralised appropriation of the agrarian commodity was underway. The literature converges on the turbulent early 1840s as its temporal stage. Scholars

83 For more detailed examples, see Owen, *Cotton*, Chapter 3; Marsot, *Egypt*, Chapter 11. The American consul in Cairo at the time, concerning the ban on local boats transferring linseed bought by European traders, claimed that the boat operators were threatened by death if they disobeyed: George R. Gliddon *No. 2: An Appeal to the Antiquaries of Europe on the Destruction of the Monuments of Egypt*, London: James Madden & Co., 1841, 151–2.

who traced closely the impacts of land redistribution on the fall of Mehmed Ali's industrialisation project only briefly mention, if at all, a cattle plague (also known as the steppe murrain, the rinderpest, or the cattle murrain) that devastated Egypt in 1841–4, and contributed, as a purely natural factor, in their view, to changing the balance of power between the Pasha and his upper class, the new landed elite.[84]

I argue, however, that the act of redistributing land was the trigger for the cattle plague's arrival to Egypt, which, in turn, pushed socioecological conditions towards ending Mehmed Ali's industrial project during the mid- to late 1840s. Once the princes, senior officials, and other dignitaries, beside the Pasha himself, were assigned new extensive estates, and with the Pasha's well-known relentlessness in pursuing and demanding the consummation of his decrees, more power/energy needed to be conjured to make the estates productive. If agricultural production fell with the abandonment of whole villages across the 1830s, with the fleeing peasants relocated to urban centres and other rural provinces, the abandoned land – now restructured as estates of *'uhdas* and *chifliks* – needed to be re-equipped with labourers, humans, and animals alike. Meanwhile, regarding the cattle population, across Mehmed Ali's reign, he was careful to have only enough cattle for powering the country's production and no more, since Egypt had no pastures for raising the cattle or other domesticated mammals. The Pasha's investment in livestock production was 'much less than his keenness on growing crops'.[85] If animal power in the Nile valley was just enough for running the apparatus of production at the turn of the decade, with the urgent demands on the newly formed landed nobility, more animals had to be quickly imported to the estates for increased productivity. Cattle were urgently needed to run the irrigation *saqiyas*, plough the land, and help corvée labourers maintain the canals and bridges.

One cannot help but assume that the cattle murrain was closely connected with a sudden leap in cattle imports to fulfil these tasks on the new estates. Fortunately, we do not need to look farther than basic

84 For more on the cattle plague, see Owen, *Middle East*, 76; Owen, *Cotton*, 62; Marsot, *Egypt*, 252. On the other hand, Fahmy and Alleaume, for example, never bring it up. Al-Hitta mentions the cattle plague of 1841–4 in a chapter dedicated to animals in the context of Egyptian agriculture under Mehmed Ali: Al-Hitta, *Tarikh*, 357.

85 Al-Hitta, *Tarikh*, 336. Alan Mikhail more recently highlighted this policy: Mikhail, 'Unleashing'.

primary sources that accounted for the 1840s cattle plague to test this hypothesis. M. Renault, the inspector-general of the French Veterinary Schools at mid-century, collected evidence and lessons learned on cattle epizootics across the world for a treatise he published in 1856 on domesticated-animal diseases. About Egypt's 1841–4 case, he started his account with the following cause of the murrain, stated matter-of-factly:

> In 1840, *the Viceroy of Egypt having divided the greater part of the territory of Egypt into appanages for his children and the high civil and military functionaries of his government* [i.e. *chifliks* and *'uhdas*], and reserving for himself vast properties, the necessity for cultivating the ground to a greater extent was felt. But the number of animals proper for labour, and above all for the irrigation of the soil, being insufficient, *it was imperative to procure more from foreign countries*. For this purpose, the Viceroy, his son Ibrahim, and several other great proprietors, sent agents into Anatolia and Karamania – countries which, by their proximity, the quality, quantity, and price of the cattle they furnish, seemed to offer the greatest advantages.[86]

Renault then explained how 'numerous purchases were made' in the winter of 1840–1, heads bought from Turkey and northern Syria. Before they reached Alexandria, however, signs of the plague showed. On the high seas, the cattle ships – transporting the animals purchased across the globe for meat and work – had temporary wooden pens that looked 'like birdcages perched on deck', where the animals were congregated in great numbers and close proximity on long marine trips. It was a big transcontinental economy that brought, among other things, epizootics.[87] It is also noteworthy that the global trade in cattle was geared towards producing cattle for meat consumption. One can expect that perhaps many of the shipments to Egypt at the time held animals raised for meat, not as motive power. Instead of halting the importation of more potentially diseased and/or 'meat' animals, more were shipped in

---

86 Cited in George Fleming, *A Manual of Veterinary Sanitary Science and Police, Vol. I*, London: Chapman & Hall, 1875, 500. Emphasis added.

87 Chris Otter, *Diet for a Large Planet: Industrial Britain, Food Systems, and World Ecology*, Chicago: Chicago University Press, 2020, 31, 42–3.

across the spring of 1841, those that died being thrown overboard on the way. Those that survived the trip landed in Alexandria in 'a generally deplorable condition'.

The process kept its momentum for the next two years, with the same kind of losses and more. The buyers, on receiving the cargo, sent their cattle off to their destinations, the estates, but, wherever they went, they spread the disease to the local animals, extending from the oxen and cows to the buffalos. With these beasts perishing, estate holders had to use horses, donkeys, mules, and camels for irrigation and ploughing.[88] Dr Ismail, an Egyptian veterinary surgeon, wrote that the newly imported oxen he saw landing in Alexandria in 1843 were of the steppe oxen breed that originated from 'southern Russia'. After giving a full descriptive account on the animals he saw, known to be generally sold in Germany, Russia, Poland, Anatolia, Karamania, and Syria, he added that he saw eighteen such animals on Ibrahim Pasha's estates a year and a half earlier (*c.*1841).[89] *The Times* on 27 October 1842 captured the losses for the year in the Delta, and how the Wali resorted to extreme emergency measures to make up for the lost animals: the Pasha sent three artillery regiments with their horses to his farms; he imported horses from northern Syria.[90]

Overall, Dr John Gamgee, the Scottish veterinary surgeon, estimated that no less than 90 per cent of oxen, cows, and buffalos in Egypt, 665,000 heads, died between 1842 and 1844.[91] In November 1845, it was estimated that, over the year, 36 per cent of all cattle used for work across the Pasha's *chifliks*, or 19,910 out of 54,972 heads, died. The Pasha's letter to the *chiflik* administrators that mentioned these numbers lamented the aggravated loss of having all these animals brought recently 'from faraway lands and from overseas'.[92] Perhaps the murrain dragged on

88 Fleming, *A Manual Vol. I*, 500–1.

89 Ibid., 501–2.

90 Cited in ibid., 499.

91 John Gamgee, *The Cattle Plague: With Official Reports of the International Veterinary Congresses Held in Hamburg, 1863, and in Vienna, 1865*, London: Robert Hardwick, 1866, 314. There is evidence in the British press that the cattle murrain was extended to parts of Britain around the same period. For example: 'The Murrain Among Cattle', *Kendal Mercury*, 17 July 1841; 'Epidemic Among Cattle', *The Scotsman*, 21 June 1845.

92 Order to *Ma'moors* of *chifliks* and *'uhda* lands dated 3 Zulqi'da 1261 H [2 November 1845], in Abbas (editor), *Al-Awamir Vol. II*, 442–3, item 3888; Sami, *Taqwim Vol. 2.2*, 535.

beyond 1844, and/or perhaps many animals brought in after the murrain died of overwork to make up for lost revenues in the earlier years.

Now, with this great reduction of animal energy on all farmed land and oxen-run factories in Egypt, we can glean yet another insight into the developments of the 1840s land redistribution processes, and the birth of the landed nobility in the mid-nineteenth century. Combining Barakat's meticulous combing through the land archives with the murrain's timeline, we get the picture that *chiflik* acquisitions – after the land redistribution triggered the initial spread of the cattle plague – might have been progressively an emergency fix for the dying-cattle crisis in the early 1840s in combination with resisting increased and sustained penetration of European trade. This dynamic, in turn, might have contributed greatly to a passive revolt against the Pasha from the *chiflik*-possessing princes and princesses, and the managers of his very own *chifliks*, as well as the *nouveau riche* senior officials and *sheikh al-balads*. Let us look at the *chiflik* records closely and see how they might help explain the progression of the power (in both senses of the word: political power and productive energy) crisis years.

The first two *chifliks* we have on record were from 1828: 10,199 acres granted to Princess Zeinab, the Pasha's daughter, in Mansoura, and another 10,102 acres from the same district went to her sister Princess Khadija.[93] One interpretation of this 22,000-acre acquisition – amounting to a little under two dozen villages with their lands, irrigation infrastructure, and mills – could be that the Pasha wanted to secure to the family, shy of confiscating the land in his name, some of the most lucrative rice production assets in the country before he embarked on his great Syrian adventures in the 1830s, or to make up for his losses in the Greek wars of the late 1820s (Mansoura being in the heart of rice lands).

After a decade without new *chiflik* records, in late June 1839 the Pasha granted two of his sons over 20,000 acres in Gharbiya.[94] In 1840, simultaneously with the first bouts of the murrain, the Pasha granted his two sons, Prince Halim and Hussain Bey, estates of 10,095 and 10,214 acres respectively, in Gharbiya. In June 1841, we see perhaps the first *chiflik*

93 Several land records converging on the size and date of acquisition of both *chifliks*, in Barakat, *Tatawur*, 76–7.

94 Ibid., 77.

the Pasha claimed explicitly to himself: a massive 36,692-acre acquisition in Gharbiya.[95] The confiscation of these greater estates was likely also spurred by the Ottoman order of 1841 that granted Ali and his family hereditary ruling rights over Egypt, as part of the settlement of the 1840 London Conference.[96] But perhaps in no less part it was motivated by the Pasha's realisation that the *'uhda* system that matured by that year – as Barakat's records on *'uhdas* show us – was not returning enough revenues to make up for the undermined monopolies. The crops of 1840 on the *'uhdas* must have been greatly affected by the murrain.

Over the next few months, the Pasha's confiscations continued, as he grabbed for himself more *chifliks*: 34,073 (thirty-seven villages) in Gharbiya in October 1841; 4,190 acres in Munufiya in March 1842; 3,244 acres in Fayoum, and 12,347 acres in Armant and other districts of Upper Egypt, in May 1842; at least 8,858 acres in Gharbiya, and 14,000 acres in Daqahliya, both in mid-August 1842 – and, on the same day, he granted his nephew 9,961 acres in Gharbiya. Then, in 1843, the Pasha confiscated the biggest ever *chifliks* declared across his reign, outside Gharbiya: 48,676 acres in Sharqiya (Al-Zankalon *chiflik*) and 51,701 acres (ninety-nine villages) in Behira, both in March 1843.[97] After two final big acquisitions in the Delta the same year, the new *chiflik* acquisitions were under the 3,000-acre limit, until 1846 – most of these newer *chiflik* acquisitions were perhaps to secure and consolidate the larger holdings acquired in earlier years – when the last 10,000-plus-acre *chiflik* decree was issued for his successor, Abbas Pasha (13,370 acres in Mahalla Kubra).[98]

Did Mehmed Ali feel by then that the murrain was radically undermining his industrial project as it killed the prime movers of the manufactories? Could he have shifted his attention, at least as a contingency plan, towards greater land to make up in cash crops for the lost industrial commodities? Did the free-trade treaty and leaked taxes push for the *'uhda* redistributions causing the murrain, then the murrain and more leaked taxes pushed for the *chifliks*, and then the murrain crisis deteriorated even more? Owen's still-unsurpassed account of the early

95 Ibid., 78.
96 Owen, *Cotton*, 61.
97 Barakat, *Tatawur*, 78–80.
98 Ibid., 80–4.

1840s presents the land redistribution schemes as causing – accelerated by the cattle murrain as a natural factor – the beginning of the end of the Pasha's agricultural and industrial project. The new land proprietors on all levels, *chiflik* and *'uhda* owners, their managers, and many other middlemen, started to resist taxes and produce demands from the Pasha on their land, especially given their losses in livestock that powered agricultural production across the board.[99]

Here, I propose to deploy an analysis of nature/society combinations to account for the overlapping and coevolving chain of causalities of this complicated power crisis. Across the board, we will see, the analysis gives up unidirectional thinking, and moves more towards a dialectical interpretation on macro and micro scales.

Across the 1810s–20s, Mehmed Ali's centralised mode of exploitation funded, and simultaneously spurred, an expansionist project unleashed on the Ottoman provinces. This, however, came at the expense of leakages from the mode: peasant uprisings, collective and large-scale flights, and sabotage. The accumulated effect showed clearly in the revenue trends of the state, which reached their peak in 1829. But in the 1830s, with the Syrian wars, peasants were fleeing the Pasha's cage en masse, pushing him towards even more draconian decrees that imposed unprecedented levels of control over the peasant's body: internal passports, planned and forced marriages to quickly hatch new hands, execution and prison sentences for self-maiming, and so on. The society/nature combination of the centralised and expansionist mode of exploitation, higher demands on labour, animals, and nature for perennial irrigation for cash crops, resulted in the depletion of the human and natural assets relentlessly required for the project, leading to a chronic tax collection crisis.

By 1838–41, the Pasha was on a collision course with the British Empire, which backed the Ottomans, fearing a bigger Russian presence in the lands of the Sultan – and potentially interfering with the British route to India – while also pushing for striking down trade barriers in the Levant, which were themselves the rock on which the Pasha's monopoly system stood. We also know that the Pasha was extremely keen on adding steam power to his industrial and military technologies, but the *nature* of his lands, Egypt and other possessions, did not provide

99 Owen, *Cotton*, 61–4.

him with coal or other assets needed for an independently built fossil empire, at a time when England was guarding her fossil-technological edge, and even if available from international markets, coal was still prohibitively expensive.

Let us call this society/nature combination thus: 'monopoly and regional expansion/coal-free nature'. It gave way to another society/nature combination: forcibly liberated trade that the Wali fixed with changing patterns of landholding, but once unleashed, it came in combination with the cattle murrain. In other words, the birth of large private estates caused by leaked tax and enforced international trade with direct producers/cattle murrain, or private estates/cattle murrain for short. I claim, next, that the cattle murrain crisis – the 'power as energy' crisis – coevolved and deteriorated with the micro-dynamics of the political power crisis between the Pasha and the new estate holders of the mid-1840s that resulted, inter alia, in the demise of the Pasha's industrial project and the early proliferation phase of steam technology in irrigation and specific cash-crops-processing industries. In short, while the enforced free trade by Britain led ultimately to the fall of Mehmed Ali's industrialisation, the local processes unleashed by this global-scale trigger – of nature/society combinations – go far in explaining how this happened. Let us look at these two processes of coevolving nature/society combinations.

The Birth of the Private Estates/Cattle Murrain Combination

With perhaps hundreds of senior officials and other dignitaries granted *'uhda* lands since the late 1830s, and several Viceroyal family members granted even greater *chifliks*, including the Pasha himself, there was a great rush to bring in new cattle for agricultural production in 1840. As we saw previously, Mehmed Ali had a policy of having in Egypt just enough livestock to sustain cash crops and other monopoly-relevant production processes, with less interest in investing in livestock, given the absence of natural grazing lands. The newly bought oxen came congregated in great numbers in boats, and as such were more vulnerable to infectious diseases. It was less likely to have the murrain originate among more spread-out and locally raised livestock with better immunity systems. The murrain went on a spree across these imported animals and started to spread inland once they landed in Alexandria. This in turn might have accelerated the murrain, as the *chifliks* needed more imported oxen.

In short, the end of the monopoly system with forcibly liberated trade inspired and/or accelerated the changing landholding patterns, giving birth to large private estates created overnight. The new large private estates caused the murrain, and the murrain ravaged 90 per cent of all animal energy in the country within three years. During these three years, the Pasha, the princes, and other lesser notables kept importing livestock to make up for the dead ones, on even greater newly acquired *chifliks*, leading to further losses in livestock and more state revenue leakage.

Concerning the *'uhda* lands, Hekekyan, an engineer and official who was very close to the Pasha, said in 1843 that his employer came down hard on their holders. Partly fearing leakage of agricultural produce from these *'uhdas* to British traders roaming the country, and partly keen on replacing his losses from the dead cattle and the chronically late taxes, he asked them that year to send monthly instalments of tax, claimed their crops before the seeds were sown, and asked for a quarter of the quotas imposed on them in currency, not produce. Those notables, ruined by the loss in imported and pre-owned native cattle that caught the disease, stirred to leave silently this whole new and potentially lucrative investment (an upper-class *tasahub*?), Hekekyan added, but the Pasha threatened to flog 'anybody who proposed to leave their farms'.[100]

The *chifliks* of the Pasha and his family, as they multiplied between 1841 and 1843, drained *'uhdas* and peasant lands that remained intact from the great redistribution. He forcibly seized peasants from villages neighbouring his large *chifliks* and brought in army units to work the estates. Overall, the administration of these suddenly confiscated huge estates was rendered inefficient, and the seized peasants, after losing their own cattle to the murrain spread by the cattle of the nobility and the Viceroyal family, and held by force away from their own possessions that paid for their own taxes, were even more motivated to embezzle and 'work as little as possible'.[101] Simultaneously, in

100 Hekekyan papers, cited in Owen, *Cotton*, 61–2.
101 Ibid., 62.

manufactories and mills that the Pasha owned, with oxen that ran the machines dead, the Pasha by 1843 had collected convicts and older and disabled men to power the machines instead of the animals. Production declined in tandem with increasing reports on suicides among the labourers who replaced oxen as motive power and deliberate arson attempts in the factories.[102]

In 1844, as the murrain was already claiming the last few thousand animals from among the 665,000 population it killed, Mehmed Ali had one of the first of his famous 'mental conditions' that many historians claimed were to be the end of him by 1848. The minister of finance, Sharif Pasha, was at a loss that season on how to report to the grand Pasha the disastrous situation for tax collection. For that year alone, not to mention earlier years' tax arrears, 14 million piastres were missing from the pre-estimated tax of 75.2 million piastres.[103] In official correspondence from the same year, the deficit of 'monies due from the districts' was said to be around 25 million piastres.[104] The minister discussed this with Ibrahim Pasha, who said that a report stating the facts was due. The report revealed that the villages – most probably remaining peasant lands from the monopoly years – could not pay their tax, and recommended to declare them cancelled.

It was said that the report was first leaked to the Pasha through one of his daughters to cushion the shock. The Pasha, however, went into a rage; he suspected theft from his son and minister of finance, calling the first a traitor and the second greedy. He told his court that he wanted these two men summoned, and that he would retire and go live the rest of his years in Mecca.[105] He sent Ibrahim a scathing letter spelling out his accusations. Ibrahim told his close circle that he was innocent, and that he would die with this letter pinned to his chest. Nubar Pasha, who was close to Ibrahim at the time, described the situation in the room when Ibrahim went over to fulfil his father's summons. Before they joined the Wali at his palace, Ibrahim instructed his personal guards to keep their sabres loose in their sheaths, then strike 'at whomever gives an order for

102 Ibid., 82.

103 Marsot, *Egypt*, 252.

104 For the year 1259 H (1843–4). Order to *mudirs* across Egypt dated 27 Jumada Thani 1260 H [13 July 1844], in Abbas (Editor), *Al-Awamir Vol. II*, 428–9, item 3833.

105 Marsot, *Egypt*, 252.

an attack'.[106] Ibrahim was taking precautions against a potential assassination attempt by his father, enraged over the unpaid taxes. Nothing of the sort transpired; the showdown between father and son did not occur. Instead, the Pasha returned to govern a few days later, after he promised Ibrahim to seize fewer peasants for his *chifliks* to make it possible to collect more tax in the future.[107]

Meanwhile, the new landed upper class in crisis – those threatened to be flogged if they deserted their estates – was covertly fomenting more trouble for the Pasha. Indirectly, Hekekyan surmised, some 'grandees' were behind a lengthy report sent by *sheikh al-balads* to the Pasha in 1844. Owen gleaned from several folios across Hekekyan's papers that the instigators behind the complaint of the sheikhs were owners of *'uhda* lands and directors of *chifliks*. Featuring prominently in the report were complaints against the sustained draining of their labour force of villages under the control of the *sheikh al-balads*, causing the deteriorating situation of tax arrears.[108]

The return of the *sheikh al-balads* and other notables across the provinces in the early 1840s, as owners of *'uhdas* and directors and middlemen in the governance of the greater *chifliks*, must have given the new class enough room for self-enrichment as well, beside their evident suffering from the leakage of labour power due to the inefficient compulsory management of the *chifliks*. Equipped with certain judicial and executive powers over their rural subjects, *sheikh al-balads* became freer in reallocating 'peasant land' to themselves in case a peasant died without heirs. They became more able to confiscate untilled pieces of land during the season – some of which might have been lands left fallow as their seized peasants were taken to plough at *chifliks* – or to claim a productive one unproductive and do the same.

Furthermore, with the new requisitions of peasants to work the *chifliks*, it becomes understandable that they might have preferred to bribe, or work gratis for, their *sheikh al-balads* to avoid claims on their time and bodies farther afield, and, in the process, generate more income and power to the local administrators. On 5 April 1844,

106 Nubar Pasha, *Muzakirat Nubar Pasha* [Arabic translation by Jarou Robert Tabaqian of: Mémoires de Nubar Pacha], Cairo: Al-Shorouq, [1983] 2009, 70–1.

107 Owen, *Cotton*, 63.

108 Ibid.

Mehmed Ali issued an order related to the continued fleeing of peasants, which might have sought to undermine the growing power of *sheikh al-balads* by collecting intelligence from the peasants on their transgressions. In the order, he said that, in spite of the harsh laws against *tasahub*, he kept hearing reports of peasants fleeing for the second time after being returned to their villages, and that 'there might be reasons pushing them to flee their homes'.[109] Therefore, he decided, they must be allowed to send complaints to governors, and if the governors did not address their grievances, they were allowed to send their complaints directly to him. After two years, a Turkish director of a *chiflik* complained that the rural notables 'have become as fat as pigs. Many of them are enormously rich from plundering the poor fellah.'[110] By 1846, Hekekyan noted that some village sheikhs were fomenting 'a passive revolt' by declining to pay their taxes and retaining their villagers away from corvée.[111]

The large estates/murrain power crisis was also the end of Mehmed Ali's industrial project. Even before the epizootic, many of the cotton factories were shut down as they lost their markets, with the army's demand on their product falling dramatically, and with the state budget strained as a result of fleeing peasants. Instead of directly administrating the factories, the Pasha, according to the Russian consul, discovered that allowing private investors to use his factories, while providing them with raw cotton, then buying their product, was more profitable.[112] The murrain power crisis came next to disastrously kill off the prime movers of the factories. In the middle of the plague, army units with their horses, convicts, and older and disabled labourers were summoned to the still-open factories instead of oxen. By 1845, the Pasha was no longer willing to keep drafting men and animals from the fields. That was the year when he told Hekekyan that his industrial plans had failed.[113] And let us remember that the confiscated *chifliks* before 1845 must have consumed most of his attention and

109 Order to Shura Council dated 17 Rabi' Awal 1260 H [5 April 1844], in Abbas (Editor), *Al-Awamir Vol. II*, 422–3, item 3811.

110 Gabriel Baer, 'The Village Shaykh in Modern Egypt', in *Studies in Islamic History and Civilisation, Vol. IX*, Jerusalem, 1961, 140.

111 Owen, *Cotton*, 63.

112 Ibid., 82.

113 Ibid., 83.

strained resources, signalling a shift towards seeking to invest more in crops and processed agricultural products.

By 1849, only two textile factories remained operational from among more than thirty, producing summer uniforms for the army. The others were abandoned, turned to other purposes, or were sold.[114] Owen, with other historians, frames this downfall as a failure of industrialisation, but the Pasha himself, around the time when he confessed the failure of his industrial policy to Hekekyan, knew that the failure was in securing fossil-fed prime movers in particular. Contrary to the common wisdom about the mid-1840s – that Mehmed Ali gave up industrialisation as a failed project – there is evidence that he saw this failure as a transient problem that would pass, and a new industrial epoch would be born on sounder foundations: locally mined cheap fossil energy. The fact that the experiment in factories ended with the mass death of the cattle used as prime movers demonstrated how vulnerable it was, as long as it could not be placed on the footing of steam power. To this Mehmed Ali turned his attention across the later years of the 1840s.

Murrain/Steam Technology Combination

On 8 April 1845, Mehmed Ali was waiting in his palace in Shubra for John Petherick, a British mining engineer he had hired to search for coal. On the day after Petherick arrived from Alexandria in Cairo, he was taken to the Pasha's palace. The transpiring exchange is most telling. The Pasha hired him to find sustainable supplies of coal for the next phase of fossil-based industrialisation that would have regained his independence from the free-trade agreement and England, had it taken off. As Petherick wrote in his memoirs:

> Addressing me, he [the Pasha] said, 'Welcome to Egypt'. I bowed. 'I have sent for you to travel in my country to search for minerals, but particularly for coal, of which I stand much in need. You have been very highly recommended to me, and I hope you will be successful in your researches. I have a manufactory at Boulak, which I wish you to see; and I want to enlarge it, *when you have found coal for me,*

114 Ibid.

that I may make guns and steam-engines, and feel myself independent of England.[115]

If we take liberty in interpreting these words, they reveal intentions that go one way only: the Pasha saw himself as a rival worthy of the British Empire. What was needed was the very subterranean factor that, his words reveal, made it such a great empire able to rein him in. Add coal to the assets of Egypt, he thought, and you will be able in a few years to acquire enough productive machinery and modern war assets to challenge London, by cloning on the Nile the very foundations of the Industrial Revolution. In the age of fossil capital, you have to possess such resources to survive independently. But it remained only a dream confined to the halls of the Shubra palace, as Petherick's account of the search revealed.

Ali added that he had also used the services of a French mining engineer who was going to do the same work as Petherick. The first to discover the coal, he told him, was to be handsomely rewarded. A few days later, Petherick was summoned to meet the Pasha again. He was briefed on a coal exploration project in the Petrified Forest, some miles south-east of Cairo.[116] The Pasha wanted him to examine it. He said, 'I wish you to go down the pit, see the bottom, and examine it very carefully; after which you will come to me at Shubra and give me your candid opinion on the probability of finding coal there. I am told that, judging from the petrification of the wood on the surface, it is sure to exist in a carbonised state underneath.'[117] He then proceeded with questions on coal mines in England: are they always very deep? How many men are employed in a shaft? Is coal in England going to last many more years?

After a two-hour ride out of Cairo, the engineer arrived at the exploration site. It went to a depth of 266 yards, but his professional opinion was that they would never find coal there. In the evening, he met

115 John Petherick, *Egypt, the Soudan and Central Africa: Explorations from Khartoum on the White Nile to the Regions of the Equator*, Edinburgh and London: William Blackwood and Sons, 1861, 2–3. Emphasis added.

116 The Petrified Forest, now a natural reserve within New Cairo to the east of Cairo, is a desert area rich in petrified tree trunks from the Oligocene (30 million years ago).

117 Petherick, *Egypt*, 5.

Mehmed Ali again. The latter inquired eagerly if he had descended into the pit. Petherick gave his opinion bluntly: there was no coal there. Desperately, the Pasha asked instead about the exact depth of the pit, and whether coal mines in England hit coal at greater depths. When Petherick told him that coal shafts in England indeed went much deeper, Ali retorted angrily, 'Then I'll sink a thousand yards!' Weeks passed. Petherick was told that he would be reporting to the minister of public works, being hired as an officer at the rank of lieutenant colonel, and that he would be free to plan exploration campaigns, with all requested men, tents, and means of transportation under his services, as much as he needed.[118]

However, the engineer's search for coal – across present-day Palestine and Jordan, then in an expedition to Persia – and that of his French rival, in Upper Egypt, did not bring news of coal. Around the same time, the Pasha met with a British official. Before he showed him samples of anthracite found through explorations conducted in or before the mid-1840s somewhere between Qina and Qusair, we have this scene drawn for us:

> In the midst of this room stood the Pasha in conversation with his physician. He saluted me cordially as I entered, and we proceeded to the window, where he sat down with his back to the garden. His complexion was excellent, the skin of his brow slightly sunburnt, but clear and healthy; his eye still remarkable for lustre and intelligence, but the cheeks and mouth much wrinkled. Coffee was served, and the Pasha began to talk of his wants, which were comprised in 'coal! coal! coal!' 'That is the one thing needful for me', said the Pasha.[119]

Short of the black carbon, a qualitatively different fossil-based second industrialisation, after the doomed first ox-man-machine episode, did not take place. Meanwhile, steam machinery for irrigation and other agricultural production processes started to proliferate weakly in the wake of the murrain, as the second industrialisation epoch that was sought to regain independence from British free trade faded away for good in 1848 as an unrealised dream in the mind of the dying Wali.

118 Ibid., 4–9.

119 Patton, *A History*, 239.

While Mehmed Ali was dreaming of and searching for extensive coal endowments to stage an industrial revolution, the landed class he had created while manoeuvring through the straits of British free trade was dreaming about steam engines for at least two other less grand purposes. First, with the oxen dead from murrain and their lost production-energy investments, richer land proprietors thought of radical alternatives for the long run, beasts of smoke and steel to replace the virus-vulnerable animals. Second, with the proliferation of free trade, agrarian producers' differentiation began to be felt, pushing the richer producers to think of labour-saving and quality-enhancing technologies for the estate.

High-quality cotton, in particular, required huge quantities of water that could be secured through steam irrigation pumps; and it also required more efficient ginning, achieved through steam machinery. Sugar of the purest kind was only attainable through steam processing technology, if its producers wanted to compete with the British and French refined sugar sold in Cairo, let alone compete on the international sugar market. Sugar cane itself, grown in Middle and Upper Egypt, also required great amounts of water if its quantity and quality were to increase – needing steam irrigation to lift the water rapidly and in great quantities onto the higher plateaus found in the southern Nile valley. In the 1840s and 1850s, however, as later chapters will show, European steam technology in agriculture was less advanced in comparison to other industrial applications (spinning, for example), and hence such machinery must have been very expensive, being during that time more in the nature of prototypes and niche experimental products, available exclusively to the richest of farmers in Egypt, who had the luxury of trial and error at the cost of great sums of money.

First, concerning the murrain effect, Abbas Pasha, beside Mehmed Ali and Ibrahim Pasha, was perhaps one of the first to consider replacing oxen with steam machinery. He spoke of this with Hekekyan in 1842, during the early stages of the cattle plague:

> His Highness observed that the introduction of steam-power for ploughing as well as for irrigation would be the only and effectual remedy for the evils which the agriculture of the country would be liable to suffer from similar epidemics and was pleased to order me to take the necessary measures to procure on his own private account

a locomotive steam-plough from England as an experiment on his own fields.[120]

We do not have supporting evidence, however, on steam-ploughing locomotives for Abbas or others that early in the decade. At the time, Hekekyan said, steam engines of this kind cost almost double the investment of using animals for ploughing.[121] Notably, Abbas's early thinking on steam technology goes against his characterisation in the literature as conservative and reactionary, generally opposed to technology and progress.[122] Perhaps he was so on many fronts, but the necessities of capital accumulation made him, as archival evidence shows, an avid enthusiast for steam technology on his estates, as much as Ibrahim, said to be 'the first Egyptian to purchase a steam pump',[123] probably for his sugar plantations in Middle Egypt.[124] One can be conservative and reactionary, and still open to invest in new technology if it serves one's immediate purposes. While some historians seem to have assumed that progressive thinking and technological progress were one and the same, Abbas simply sought technology to increase profits.

Around the same time, the grand Pasha himself had fixed his interest on securing summer irrigation through the indirect use of steam engines in the Delta by the mid-1840s. After seeing with his own eyes how efficient the steam pumps of the renewed Alexandria docks were in pumping out sea water, he ordered the engineer who oversaw the works, as they both sat during the inauguration ceremony of the revamped docks, to start the construction of the long-discussed Delta barrages. This meant requesting, for pumping water out of zones where construction was to proceed, 'four pumping engines of 100-horse power each' from France, in December 1844.[125] The claimed feasibility of the planned project was extended to replacing with direct canal irrigation the work of 200,000 oxen and half that number in men at the *saqiyas*.[126] The Delta

120 Owen, *Cotton*, 64. We will see later that successful steam-ploughing engines were only commercially produced in the early 1860s (Chapter 7).

121 Ibid.

122 Marsot, *Egypt*, 253–4.

123 Owen, *Cotton*, 64.

124 Fowler, *Report*, 15.

125 'From Alexandria, 6 December', *Shipping and Mercantile Gazette*, 31 December 1844.

126 Ibid.

barrages were first planned in 1833, before Mehmed Ali in 1844 reassigned the project to Mughel Bey, the engineer of the renewed docks. It was not completed, however, until 1861.[127]

Besides these four big steam pumps, Mehmed Ali may have spoken of his wishes for replacing oxen/*saqiyas* with steam pumps, in the middle of the devastating epizootic, in early 1843. An anonymous reporter from Alexandria to the *Evening Star* wrote:

> The most striking occurrences of this period in the history of Mahomed [*sic*] Ali's life, is the sad mortality among the cattle of the country; there has been a loss of nearly 300,000. You will understand that the main use of them has been for irrigation. The great question that now engages the mind of that amazing man is not only to make up for the sudden loss, but to anticipate another such dispensation. The great projects that he has now in view with reference to damming up the hills [typo for 'Nile'?], are highly important, and, if well devised, no doubt would effect the desired object. *But what strikes him most as the immediate remedy for the sudden loss of the cattle, is a judicious application of irrigation by means of steam*. The Messrs. Galloway have long devoted their attention to this important subject; and it is stated that they are prepared with a plan by which the irrigation of the country can be effected at 400 per cent less than the present mode. [. . .] If ever there was a country more susceptible than another for such improvement, it is Egypt.[128]

These plans, however, did not materialise any time before the early 1850s (see Chapter 3). If anything, the epizootic set the eyes of the nobility on long-term steam substitutes for oxen on the farm. It brought with it no grand plans of industrialisation, and when it flowed, it helped bring down Mehmed Ali's factories for good. When it ebbed, it incentivised investments in steam technology for producing cash crops, to be processed through factories elsewhere. Moreover, concerning the impact of the land redistributions – the process that started and coevolved with the progress of the cattle plague – on early proliferation

127 See Jennifer Derr, *The Lived Nile: Environment, Disease, and Material Colonial Economy in Egypt*, Redwood City, CA: Stanford University Press, 2019, Chapter 1.

128 'Extract from a Private Letter', *Evening Star*, 1 February 1843. Emphasis added.

of steam technology in agriculture, we have evidence from two commodities: cotton and sugar. Here, we stop only at the evidence put forward by cotton, as progress in steaming sugar production will be discussed in the next chapter.

By early 1844, we see the first evidence, from the *Shipping and Mercantile Gazette*, that cotton was starting to flow profusely through the fingers of the Pasha at the docks in Alexandria, as his monopoly-replacement tactics were failing against mounting free-trade pressures:

> The effects of free trade in Egypt have been strongly felt by the great activity which has existed in our [i.e. Alexandria's] cotton market during the last three weeks [prior to 20 March 1844]. This staple has been arriving here [Alexandria] *in numerous small parcels, belonging to private individuals*, and is eagerly bought by the merchants.[129]

As the cotton crop came through, the owners of the larger estates managed to produce cotton of the highest quality, fetching highest prices at the docks, while smaller farmers devolved towards selling lower qualities. The production of high-quality *Jumel* cotton benefited from rotating the crop, once every four years on any given piece of land, on larger estates, where the owner had the luxury of moving the crop between his many plots. It needed more irrigation, while lower-quality cotton was watered only twice per year, following the rhythm of the inundation. The *chifliks* were some of the most privileged lands in terms of the availability of water for irrigation. Finally, the higher-quality and more profitable cotton that the large estate holders, royal family, and others set their eyes on required the use of steam technology in ginning.[130] As early as 1845, a Khurshid Pasha was planning to replace his thirty-two oxen-driven cotton gins – processing his 30,000-acre cotton-growing estate in the Delta – with two 14 hp steam-ginning machines.[131]

We will see in the next chapter how steam technology and energy continued to weakly proliferate across the 1850s to cotton and sugar production, reaching an unprecedented level in cotton-ginning steam

129 'Egyptian Commerce: Alexandria, 20 March', *Shipping and Mercantile Gazette*, 6 April 1844. Emphasis added.

130 Owen, *Cotton*, 71–81.

131 Ibid., 64.

technology by 1862, but, for now, we need to stop and take stock of the progress of the story so far.

Conclusion: 'Coal! coal! coal! That is the one thing needful for me'

In this chapter, reinterpreting the last two decades of the Pasha's reign from the viewpoint of steam power and fossil technology, I sought to explain this history within the framing of a multifaceted power crisis, of energy and politics. After plotting it out in detail, in the following I present two notes focused on macro-dynamics: the first on the build-up of this crisis towards its climax in the mid-1840s, and the second on its repercussions over the following decade and the trajectories that ensued towards the 1880s.

The power crisis framing is inspired by the cattle epizootic of the early 1840s. Investigating it in a way that recognises nature as coevolving with the social in dialectical combinations gave rise to insights on several fronts. We have seen the formation of the landed-nobility class as born out of and within a state-wide catastrophe that annihilated animal motive power on these very lands; the undermining of Mehmed Ali's industrial project, which was perpetually waiting for coal to be discovered, and the dying oxen that let it down before that happened; and the earliest phase of deploying steam power in Egypt for agrarian commodity production in the aftermath of the crisis.

Looking at the bigger picture in which this power crisis was embedded, we can see the broader and more extensive dynamics and structures that preceded it. Domestically, Mehmed Ali's coal-deprived industrialisation and cash-cropping for international trade exhausted the rural population, giving rise to a multitude of local revolts across the 1820s and seemingly systemic processes of fleeing taxation, conscription, and corvée impositions, through wide-scale *tasahub* and self-maiming – to purposefully displace oneself or debilitate parts of the body or its senses to avoid even worse fates. The broader domestic context seems, then, to have been one of a bigger and deeper crisis of manpower, one of tragic proportions for the lives of millions of rural subjects over decades. It culminated in a chronic tax collection crisis starting in the early 1830s, around the time when the Pasha took a 100,000-strong army of the same peasants to wrestle the Levant from the hands of its Emirs and dig for coal, and develop the region the same way he did Egypt – all in the hope, perhaps, of reaching a balance where coal mines would make the self-inflected plight of the peasants redundant and erase the word *tasahub* from the dictionaries. In association with coal prospecting, we saw from his own words with European friends and in official correspondence that it was desired as a means to free him from England. But he also might have been hopeful that machines, if deployed en masse after finding coal, would, again in his own words, 'save man from types of toil he was not created for'.[132]

Here is a Pasha whose every effort to hit abundant coal seams to make guns and steam engines in order to 'feel [himself] independent of England', as he told geologist John Petherick, came to a dead end. Around the same time he expressed these carboniferous hopes for potentially wriggling away from England's fossil-empowered global reach – only three years before his death in 1848 and a few years after England defeated him in the Levant and brought free trade to his extremely exploited Nile valley by imperial naval force – Mehmed Ali showed another British friend anthracite samples found in Upper Egypt and exclaimed in frustration, 'Coal! coal! coal! That is the one thing needful for me.'[133]

What exactly was he thinking during the ongoing frustrations of his quest for locally sourced coal? Was he racked with envy of England, the country that burned most of it? He must have seen himself and the possessions under his hands as a rival worthy of the British Empire. He had requested steam engines from England in the early 1820s – when they were still restricted to England – and was turned down, so he brought them from France at an inflated price. He brought in the great Galloways to run his factories' machinery during the same decade and built himself a great modern (but coal-deprived) foundry. There, he used blueprints of steam-powered spinning machines like those of England to forge his own steam-powered machinery and powered them

132 Ali El-Gritly, *Tarikh Al-sina'a fi Al-nisf Al-awal min Al-qarn Al-tasi'a A'shar [History of Industrialisation in the First Half of the Nineteenth Century]*, Cairo: Dar Al-Ma'arif, 1952, 87–8, citing correspondence from *Diwan al-Fabriqat* (Manufactories Department), 14 Safar 1257 H [4 April 1841].

133 Patton, *A History*, 239.

instead by oxen. One can read his motives on acquiring coal as those of a practical man who considered it his destiny to compete with England, but only once he had found coal endowments equal to English endowments. And he wanted to know, as he asked his hired engineer Petherick in 1845, how long is coal going to last in England?

In other words, the British Empire was likely the rival he was building a mini-empire to match and stand his ground against, independent from it, free to expand territorially and keep building power and commercial might – not his Ottoman overseers, whom he defeated repeatedly across the 1830s over the plains of Greater Syria. And it was England that brought down his coal-frustrated empire in direct and indirect ways. Besides the defeat in Acre by British naval steamers and the eventual undermining of his entire Syria adventure, it was British coal endowments that indirectly co-shaped his commodity-production endeavours – the then-emerging British fossil empire.[134] British demand for grains and cotton had a major stake in his efforts to turn to perennial irrigation (with the substantial infrastructural work that entailed) to grow cotton extensively, and start industrialisation while he searched for coal to feed it.

Let us look at another crucial aspect of the impacts of Britain and its trade: the formation of the Egyptian landed class at the eye of the storm of the early 1840s power crisis of the cattle epizootic. It would not have been formed suddenly in that particular way – the engineering of a land-wealthy class overnight, given land and associated tax responsibilities, and above all hoped-for loyalty to keep the commodity monopoly intact – had it not been for the free trade Britain sought to impose on the Pasha's lands, in combination with mounting tax losses as the peasants kept fleeing the land. The fix that he came up with was conducted against imminent British military threats as much as it was against imminent British trade interests in raw material to feed the growing industrial capitalism. This fix was the result of a manpower that had been thoroughly depleted over the decades. It aimed to feed large-scale trade, industry, and military power that emerged with industrial capitalism. It fed it raw material, tried to emulate it in principle – once the abundant raw material showed up and looked inviting to process through

134 Andreas Malm, *Fossil Capital: The Rise of Steam Power and the Roots of Global Warming*, London: Verso, 2016, 267.

factories – and protected the whole enterprise with regional expansion. Growth without perceiving of limits brought the crisis, and the crisis occurred when the productive powers that created this growth hit their material limits.

The power crisis, in its global context, can then be seen in how British fossil capital, in its global circulation (as it took the form of a fossil empire), rubbed up against an ambitious and aspiring exploiter who fed this global circulation, but, simultaneously, stood his ground with temporary measures for powering a grand-scale industrial adventure that Britain did not think useful to her interests, while he continued to frantically seek the very material that powered this superior industrial capitalism of England and growing global might: coal.

Let us then venture to entertain a concluding thought: Mehmed Ali's reign, not just the 1830s and 1840s, was a protracted power crisis. A power crisis that plagued an ambitious endeavour for a certain centralised state formation, enabled and simultaneously undermined – as an independent and predatory formation – by the growing British fossil empire. Modern Egypt, if we equate modernity here with the rise of the Capitalocene, or the act of igniting the engines on the Nile and the raising of chimneys on its horizons in a scene matching the rising industrial centres of England at the time, was born out of this protracted crisis of the nineteenth century's first half.

It was during those decades that Britain's fossil empire would not let a hopeful rival have his way with a crucial and extensive section of its route to India. The landed class this rival engineered to resist the march of free trade in raw material for England's steam engines became the force that led the deployment of fossil-fuelled technology in Egypt over the later decades of the century, starting in the last few years of Mehmed Ali's reign.

In sum, if we are to take a bird's-eye view of this process, it is that of a failed upwards curve of independent mass commodity production and trade, followed by a longer downwards curve of growing dependence on the core of industrial capitalism until the 1880s, the point at which precolonial Egypt ceased to exist. Just after that moment, at the turn of the twentieth century, a British cabinet minister said something that seems to describe retroactively England's friction with Mehmed Ali's crisis-plagued predation from the 1830s and '40s: 'we are like an octopus with gigantic feelers stretching out over the habitable world, constantly interrupting and preventing foreign nations from doing that which we in the past have done ourselves'.[135]

The rest of Part I, then most of Part II, will take us forward across those decades between the 1840s and the 1880s. For now, it is sufficient to highlight how the newly formed landed class in the 1840s, as we gleaned from the last section in this chapter, started deploying steam engines for irrigation, cotton ginning, and sugar refining, partly to remedy deteriorating maintenance work for irrigation infrastructure over the 1840s and 1850s, and partly dancing to the tunes of changing global standards and expectations of the quality of two of the commodities produced from cash crops in Egypt, sugar and cotton. Chapter 3 eases us into the early 1850s.

135 Lord George Hamilton to Curzon, 2 November 1899, cited in Darwin, *The Empire*, 66.

3

The Sugar Factory, the Cornish Pump, the Mud Machine (1848–55)

Then the modern landscape begins to stir. It is foggy; the black smoke of our machines can hardly be distinguished from the grey, drifting clouds; our engine whistles, and behind the forest a second whistle answers; not the echo, but another locomobile slowly pulling an instrument even stranger than ours over the stubble. There they drain with steam; also an invention of our brilliant Fowlers! Five minutes pass and through the morning stillness, which is not yet broken by the soft whirring of our cultivators' belts, a third whistle sounds far away from the next farm; there a steam engine sets in motion, throwing mountains of threshed straw about it. This is how the landscapes of the future will probably be shaped everywhere.

Max Eyth, letter from Hertfordshire, 30 October 1861

What *Pashmuhandis* Eyth Did Not See

Here, and in later chapters, we look at the early career of the engineer Max Eyth, after the cattle plague and the power crisis of the 1840s, in order to examine the contours of Egypt's transition from oxen to steam engines. His memoirs, letters, and other books – including poetry and a novel, no less informing – besides offering an extremely important testimony on the zeitgeist of progress in agricultural technology in the third quarter of the nineteenth century, present information rarely tapped into before in histories on Egypt.[1]

Born to a Swabian evangelical family, his mother and father also writers, Max Eyth (1836–1906) studied mechanical engineering at Stuttgart before joining the Fowlers of Leeds (who pioneered the design and production of portable steam-ploughing engines) in 1861 as an apprentice in agricultural machinery. He kicked off his career in Egypt as the representative of Fowler and the holder of the title chief engineer (*Pashmuhandis*) of Prince Halim, Khedive Ismail's uncle. He then travelled across the world for twenty years, installing agricultural steam machinery across continents, before resettling in Germany in 1882. In 1885, he founded the German Agricultural Society and was its first director.

In 1861, once Fowler recruited him, he prepared to travel to Egypt to set up two steam ploughs commissioned for Prince Halim at his farm in Shubra, a few miles north of Cairo. The Prince requested in a later telegraph, however, that the Fowler representative coming with the machinery should stay for a year. Since Fowler needed Eyth at the London International Exhibition of 1862, another engineer was sent to Halim's Shubra works instead.[2]

On 13 January 1863, days before Ismail became the Viceroy, Eyth was waiting at the docks to learn which steamer he was going to take to Alexandria. For Prince Halim had requested another engineer from Fowler, to come and sort out problems encountered with the two steam cultivators purchased the year before. The cotton boom of the early 1860s in Egypt was on Eyth's mind as he sat waiting for one last telegraph telling him which steamer to board: 'This Augean stable [i.e. Halim's machinery in disarray] should now be swept out as quickly as possible; for the world cries out for cotton, and the cotton for the ploughed land.'[3] If Eyth, at that point, thought that his employer's machinery was the first ever to touch Egyptian soil, he soon learned otherwise. Fowler's two ploughs were

1 I have Rosa Luxemburg's citation of Eyth in *Accumulation of Capital* (Chapter 30) to thank for the rediscovery of this important stream of primary sources. Eyth's technical treatise on steam cultivation in Egypt also came up in David Landes's monograph *Bankers and Pashas*.

2 Max Eyth, *Im Strom unsrer Zeit. Erster Band: Lehrjahre* [In the Flow of Our Times, Vol. I: Apprenticeship Years], Projekt Gutenberg-DE, 1904, Chapter 22 (letter dated 29 November 1861).

3 Ibid., Chapter 41 (Leeds, 13 January 1863).

preceded, as he saw in February 1863 at Shubra, on a tour with the Prince, by around twenty other steam engines – at Halim's Shubra estates alone. These were steam engines for ploughing, pumping irrigation water, cotton-ginning, and sugar-cane crushing.[4]

In all likelihood, however, Halim's great assemblage of twenty steam engines in 1863 at Shubra, not to speak of his other estates across the country, most notably in Muta'na in Upper Egypt and Kafr Al-Sheikh in the northern Delta, was a rarity at the time. In fact, Halim might have been the first person in Egypt to introduce steam engines to agriculture on this large scale, but he was not the first in the Viceroyal family to bring the technology. Although our engineer did not provide information in detail on the pre-Halim machines (we turn to that shortly), he periodised those early steam engine undertakings as spanning 1850–62. In his opinion, if not for the colossal sizes of the grand Pashas' estates, and the power they wielded to facilitate the introduction of the new technology, it would not have happened. Even if he was speaking of seedling efforts in that direction by Abbas and Sa'ed Pashas, he also included Halim's steam endeavours as central to the steam technology's earliest phase on the Nile. Being the youngest surviving son of Mehmed Ali, a great agriculturalist, and a strong supporter of technological progress, as Eyth learned across a few years of working closely with the Prince, he became 'an unwavering champion for the introduction of machines', even if Ismail Pasha 'followed him in a quantitatively larger scale' after the spring of 1863.[5]

Eyth was not present in Egypt between 1850 and 1862 as a witness to those earlier steam engines, so one can suspect that his assertions on how steam technology took off from Halim's plantations in the early 1860s are incomplete. Meanwhile, the few histories that investigated prime movers in nineteenth-century Egypt either stopped at the age of Mehmed Ali or glossed over the 1850s and 1860s, before asserting that the mechanisation of agriculture started with the steam pumps on large estates in the 1870s.[6]

4 Ibid., Chapter 43 (Cairo, 21 February 1863).

5 Max Eyth, *Das Agricultur-Maschinenwesen in Aegypten [Agricultural Machinery in Egypt]*, Stuttgart: J. B. Metzler'sche Buchhandlung, 1867, 22.

6 Ali El-Gritly, *Tarikh Al-sina'a fi Al-nisf Al-awal min Al-qarn Al-taṣi'a A'shar* [History of Industrialisation in the First Half of the Nineteenth Century], Cairo: Dar Al-Ma'arif, 1952. Ghislaine Alleaume, 'An Industrial Revolution in Agriculture? Some Observations on the Evolution of Rural Egypt in the Nineteenth Century', in *Agriculture*

So, beside engineer Eyth, we turn next to those who witnessed the steam machinery in action in the 1850s. As we move to relevant accounts from travellers to Egypt at the turn of that decade, we see yet another face of the story beside the Prince Halim–centred narrative Eyth provided, one that pinpoints the imperatives of the refined sugar market beside the few cotton-centred steam-pumping advancements in the Delta. In the lands of sugar, Eyth's account of foundering steam gives way to fully developed steam technology, even if it was concentrated in a few factories.

Early 1850s: Steam for White Sugar

When Bayard Taylor, an American poet and diplomat, was in Egypt in 1851, he visited in Reyremoun, Minya, the factory that sources claimed was Mehmed Ali's first big sugar manufactory, built in 1818 with the help of an English engineer. There, Taylor saw 'a large manufactory' where the sugar processed by the peasants in other more primitive establishments was further refined and whitened before it was sent to Cairo.[7] It was the property of Ibrahim Pasha before he died in 1849, said to be the first person to introduce a steam pump to the country.[8] A commercial travel guide to Egypt from 1867 supported this claim about the Reyremoun facility's specialisation in refining peasant-manufactured sugar, a seemingly extensive output system. The factory, according to the guide, produced white and brown sugar. During the 1860s the white sugar was sold in Cairo as high-quality sugar, while the brown, coarser variety was sold in Sudan and farther south.[9] It seems that, in that factory at that time, only steam-powered refining machines were used, and not crushing mills. Although Taylor's account is not conclusive on the prime movers used, the travel guide described the cane-crushing process as powered by oxen, but refined white sugar was generally refined through steam-powered

in Egypt: From Pharaonic to Modern Times, edited by Alan Bowman and Eugene Rogan, Proceedings of the British Academy, Vol. 96, 1999, 338–9.

7 Bayard Taylor, *A Journey to Central Africa, or Life and Landscape from Egypt to the Negro Kingdoms of the White Nile*, New York: G. P. Putnam (10th Edition), 1859, 104.

8 Thomas Fowler, *Report on the Cultivation of Cotton in Egypt*, Manchester: J. J. Sales Printers, 1861, 15.

9 Gardner Wilkinson, *A Handbook for Travellers in Egypt*, London: John Murray, 1867, 289.

machinery. This disparity between prime movers in crushing and refining machinery is consistent with the claim that the factory was refining and whitening sugar produced elsewhere, since the factory's oxen-led mills were not responsible for producing all the raw sugar gorged by the faster-paced steam refiners.

But even before Taylor's journey, full steam technology for sugar existed in 1850. William Arnold Broomfield, a British botanist, sent letters to his sister about his travels in Egypt, including his visit to a sugar factory near Minya, owned by Prince Ilhami, the son of Abbas Pasha. There, 'the evaporation of the cane juice is carried on in vacuo, as in the great English refineries'. 'The machinery is of the most complete description, and consists not only of the sugar mills, etcetera, required for refining, but of powerful English steam engines for driving them, and pumps of great calibre for irrigating the land of which 1500 acres are planted with canes.'[10] It is noteworthy here that Max Eyth later categorised a similar 100 hp steam pump he saw in the Delta as a 'Cornish pump' (from Cornwall, England) – the beam steam engine used earlier to pump water out of coal mines.[11] Close by, Broomfield saw, at the Rhoda factory, owned by the young Ismail at the time, 'a steam pump, etc, for irrigation, and others in course of erection'. From Rhoda, the site of an extensive operation at the time, the sugar produced was sent to Cairo, where an English firm took care of its refining before it was sold.[12] Further south, around Ernout (Armant), Broomfield saw immense sugarworks in construction said to be the property of the young Mustapha Bey, Ibrahim's son, beside Ismail and Ahmed. The 'huge chimneys of the engine and boiling houses' were nearing completion that winter of 1850.[13]

By 1850, in Middle and Upper Egypt, at least two factories were fully or partly operating on steam technology, with beam engines originally developed for pumping water out of coal mines, to irrigate adjacent sugar cane plantations, and a few other similarly steam-powered

10 W. A. Broomfield, *Letters from Egypt and Syria*, London: William Pamplin, 1856, 70–1.

11 Eyth, *Das Agricultur-Maschinenwesen*, 20.

12 Broomfield, *Letters*, 70–1. Perhaps Broomfield's reference to an English firm for refining sugar was to a Cairo workshop where the refined product was packed in the blue packaging said to be imported from England to give the sugar the appearance of British imported sugar.

13 Ibid., 153.

establishments were being erected. When did these engines make it to Middle Egypt if some of them were already in full operation by 1850?

The process might have started in the mid-1840s, with Ibrahim Pasha's visit to the factories of Birmingham, Manchester, and Belfast in the summer of 1846, widely celebrated by the British press. It was two years after the tense loose-daggers showdown with his father at Shubra palace, and months after Mehmed Ali's search for native coal with the help of French and British engineers brought bleak news and he confessed the failure of his industrialisation project to his engineer Hekekyan. With the redistribution of land, and in the wake of the devastating cattle murrain, Ibrahim was dreaming of steam engines, coal, and schools for raising technicians for maintaining the machinery and disciplining the hands. Was he contemplating a great Egypt-wide industrial renaissance, or just caring for his own estates, especially the sugar cane plantations and factories of Middle Egypt? The evidence at hand does not tell. It might be interpreted as indicating wishes somewhere on a spectrum between this and that, but it shows more solidly how steam engines on the sugar mills and on the estates for irrigation might have started to show up in the early 1850s.

On 18 June 1846, Ibrahim, with his translator Nubar and a few others, started his factories tour of England at the King Edward Free Grammar School in Birmingham. Among his many questions to the headmaster was whether they admitted foreign students. He was told that this could be arranged for tuition fees. He was greatly interested in exploring the educational facilities as much as the factories, for he reportedly wrote back to his father saying that 'the superiority of the Frank [*Frinja* in Arabic, usually used to connote European] civilisation was owing to the better conditions of the humbler classes'.[14] He proceeded to Birmingham's 'Soho Works, with which the name of Mr Watt and the steam-engine are so intimately associated', where 'every species of machinery which ingenuity could devise for the improvement of art readily found patrons in distinguished proprietors':[15] the owners of the great foundry that produced steam engines since it was

14 Andrew A. Patton, *A History of the Egyptian Revolution: From Arab and European Memoirs, Oral Tradition, and Local Researches, Vol. II*, London: Trubner & Co., 1863, 239.

15 'Ibrahim Pacha', *Saint James's Chronicle*, 18 June 1846.

founded in 1775 by Matthew Boulton and James Watt themselves. Upon his arrival at the Soho Works,

> his Highness was conducted through the principal workshops, and his attention was very soon arrested by the engine department, in which are some specimens of the most admirable workmanship. *He was also shown engines similar to a pair which were made in this establishment last year for his father*, and which were about 150-horse power each [must have been beam engines for irrigation, given the great hp]. These engines seemed to please him very much, and after examining the different sorts exhibited before him, and ascertaining the prices of each, *he gave orders for eight engines for pumping water, and other purposes*. He next examined several immense boilers [thinking of upgrading his Middle Egypt sugar factories?], and paid particular attention to the various processes of manufacturing through which they were put.[16]

When he was introduced to the enormous 'pattern rooms' where machines were manufacturing steam engines, he reportedly expressed his awe at the sight, saying that now he understood why Britain 'excelled the world in [her] manufactures', and that 'it was out of the question to talk of competing with England while [she] had places like Soho [Works]'.[17] After purchasing those eight steam engines for irrigation and other purposes, and after seeing the likes of the two ordered engines in 1845 by his father, he visited a big glass factory. There, 1,500 to 2,000 workers laboured every day. He admired the simplicity of the production process run by steam engines and expressed admiration at the 'moral and religious care' given to the many child workers at the factory. They reportedly had schools where 'they were all taught useful knowledge'.[18] Did he visit the glass factory to see how he could imitate it on the Nile? Or, more practically, to see with his own eyes how the boilers applied great and sudden bursts of heat energy to the material, not unlike how the boilers of sugar factories should optimally operate?

16 Ibrahim's visit to Birmingham: 'Ibrahim Pacha', *Saint James's Chronicle*, 18 June 1846. Emphases added.

17 Ibid.

18 Ibid.

Again, we cannot know his intentions, so let us leave it at that, and hope to glean more as we proceed with him on the visit.

Ibrahim's next stop was at 'extensive coal works belonging to Messrs. Hunts, of the Braids', in the Black Country, coal mines that sustained the Industrial Revolution as it took off some decades earlier. It was a long-sought-after moment. Since his arrival in Birmingham, he had expressed his wish repeatedly to go down a coal shaft. At the mouth of the pit, and when he was told that it went down as deep as 260 yards, and realised that the means of descending was 'a frail skid suspended by a rope', he got cold feet, and handed the rope to his entourage instead.[19] Nubar, his translator, was among those who went down. And 'while they were being lowered, and as long as he could make himself heard, his Highness continued speaking to them in his own language and apparently congratulating himself upon being upon *terra firma*'.[20] After they came up again, he was gifted a piece of coal from the mine, a 'black diamond' that fuelled industrial wonders. He apparently asked dismissively about the cost of the coal, but left it at that without making purchases, content with the new knowledge he had gleaned from his shouted questions to Nubar as he descended on the skid.[21] He soon left for Manchester.

The first place Ibrahim visited in Manchester was 'one of the largest spinning establishments in the Kingdom' at the time, the factory of Messrs M'Connel and Co. cotton spinners.[22] He only visited the latest production line at the factory, and spent forty-five minutes there, where raw cotton, including that imported from his estates in Egypt, passed through the different processes of drawing, roving, stretching, spinning, and reeling.[23] He was interested, upon spotting some of the Egyptian cotton in the rooms, in checking how it compared to the American Sea Islands variety. He then proceeded to the cloth-printing establishment of Messrs Thos Hoyle and Sons. Across the visit, he seemed most interested in how the press rollers through which the printed cloth moved for drying were cleaned. Perhaps he was also thinking of steam-powered mills in his factory at Minya for squeezing the juice more efficiently out of the sugar cane? More attention was spent on checking how efficiently

19 Ibid.

20 Ibid.

21 Ibid.

22 'Ibrahim Pacha in Manchester', *Manchester Times*, 20 June 1846.

23 Ibid.

rollers dried indigo-dyed cloth – indigo was also imported from Egypt – at the same visit. 'By way of testing the efficiency of the squeezing machine he got one of his attendants to take hold of one end of a piece of cloth from it, and assisted him to wring it . . . but without extracting, as may well be imagined, one drop of moisture from it!'[24] Did Ibrahim at that moment have a mental image of a very dry sugar cane shoot coming out of this same or adapted roller with more juice squeezed out of it to produce more sugar from the same crop?

Another piece of steam machinery that also caught his attention was a recent invention for drying cloth by centrifugal motion – the same kind of motion that produced refined sugar of the whitest variety a few years later. While the reporter was preoccupied with conveying to his readers that dyeing cloth in Egypt went back to the times of the great Pharaohs, since it was a tour for the oriental prince from that exotic historical land, Ibrahim might have been contemplating how such modern processes could be copied later on to comparable factories, or, more probably, how they seemed open for adaptation to a more efficient and more refined white sugar in his factories. However, at his next textile factory, at the cotton mill of Messrs Birley, he might have covertly expressed his frustration at the growing scarcity of 'hands' on his lands that seemed to stand in the way of working his factories and those of his father. The machine 'that seemed most to interest the prince [. . .] was one for folding the cloth'. It was able to fold ninety yards of finished cloth per minute. He reportedly told his entourage that 'it seemed to have a soul', adding, 'if a machine could be invented for making men with the same rapidity there would be no need for women'.[25]

Perhaps Nubar had enough wit to intentionally mistranslate Ibrahim's last words in a direction more expected from an oriental, to elicit laughs instead of worries about hidden agendas or hopes. Perhaps Ibrahim's tongue slipped in the wrong place and time. Perhaps he actually meant that if such a machine could make men so rapidly, he and his father would not have needed to go through the trouble of disciplining the peasants into reproduction and wait for years until their children could be driven to coerced work, or keep issuing passports and imposing great penalties on those who maimed themselves, or drowning in canals their

24 Ibid.

25 Ibid., as well as all information on the Manchester visit.

mothers who maimed them to spare them a worse fate. Perhaps he was dreaming of such a machine, with a soul, that could hatch men more rapidly than men fled their tax and corvée. But this is only guesswork. So let us stop at that and return to the early 1850s, now better equipped than before to understand how it looked and felt for the great men of Egypt before the turn of the decade vis-à-vis steam engines.

Two other American tourists provided more details on steam in early-1850s Middle Egypt, in the direction of giving reasons for why Ibrahim was not interested in purchasing coal for his sugar factories during his 1846 trip to England. An 1853 trip by Joseph Thompson and Thomas Upham, a pastor and a moral philosopher, was documented in two separate books.[26] Thompson, with Upham and others, visited a sugar factory in Minya, the property of Abbas Pasha. Upham confirmed in his book that the factory was powered by steam. At the time of the visit, it was under inspection by a certain Bey who came from Cairo on behalf of the Pasha. He told the American travellers that he was inspecting, in the south, 'the manufacturing establishments'.[27] Thompson added some more detailed information. The factory had two great hexagonal brick chimneys, landmarks of Minya back then. The sugar it produced – fine-grained sugar and 'pellucid rock candy' – was refined through French machinery. The adjacent fields were irrigated by a big steam pump, a Cornish beam engine in all likelihood. The management was French. However, the most important observation we get from the two gentlemen was that the fuel feeding the steam engines at the factory and at the irrigation stations was dried sugar cane refuse mixed with dry cornstalks: 'After the cane is pressed, it is dried in the sun and used for fuel, and this with the addition of dry cornstalks, suffices to feed the engine.'[28]

We will turn to the significance of this process of using alternative non-fossilised fuels for steam engines in Egypt later (Chapter 9). For now, let it suffice to see how the presence of other fuels besides the expensive imported coal might have incentivised the rapid proliferation of steam machinery in the then-small sugar industry, relying on fuel, at

26 Thompson dedicated his book to his travel companions, including 'Prof. Thomas C. Upham'.

27 Thomas Upham, *Letters Aesthetic, Social and Moral, Written from Europe, Egypt, and Palestine*, Brunswick: J. Griffin Press, 1855, 272–3.

28 Joseph Thompson, *Photographic Views of Egypt, Past and Present*, Boston: John P. Jewett and Co., 1854, 90–2.

least partly, sourced from the sugar cane plantation. More important, of course, was the need for steam machinery for producing white-grained sugar that could be sold on markets, local and international, to compete with the refined white sugar of Europe, after the end of a monopoly that could have made sugar production in earlier decades profitable on local markets. In other words, if the international sugar market pushed for the proliferation of steam sugar factories since the early 1850s in Egypt, the presence of cheaper alternative fuels – besides the expensive coal that needed to be imported from England – made the process even more financially viable.

Overall, the steaming of sugar factories and sugar cane plantations in Middle Egypt at the turn of the decade must have been a success story, as some factors for success were present: English and French engineers and staff for maintaining the engines, and potentially cheaper local non-fossil fuel (from sugar cane refuse, or bagasse) besides imported coal. Did this success spark the proliferation of steam pumps and other steam engines in the Delta at the time? We do not know how far the two affairs were related, but we know that the Cornish/beam engine made it to the Delta around the same time it showed up in Minya, or slightly after. In the latter case, the suspected motives were different, being largely direct continuities from the murrain/large estates combination of the late 1840s.

Early 1850s: The Mud Machine, the Cornish Pump

Bayard Taylor, before he arrived at Minya, was somewhere in the north-western Delta. There, 'in two or three places, we passed mud machines, driven by steam, for the purpose of cleaning the [Mahmoudiya] canal' on which his boat was travelling to Cairo.[29] Some years later, engineer Eyth meticulously described and illustrated a potentially different canal-dredging steam machine, saying (he was likely wrong, given Taylor's observations) that it was the first canal-digging piece of steam technology in Egypt, imported by Halim Pasha from Gargan et Fils of France, and installed at his Shubra estates. With an intricate body of cogs and belts, it consumed a lot of coal and had many vulnerable wooden parts

29 Taylor, *A Journey*, 24.

that needed maintenance and frequent replacement as it operated exclusively in the outdoors.[30] Had Eyth seen the Mahmoudiya canal's two or three dredgers from Abbas's time, he would probably have given the same verdict: the machines were inefficient, but 'the idea is a quite practical one' that will deliver better results in the future. 'Such inventions require patience, energy and money to become viable.'[31] These mud machines fought a losing battle, as by the mid-1850s the Mahmoudiya, being a prominent example indicative of the state of other Delta canals, was on the verge of total collapse, to the extent that it needed to be re-dredged nearly from scratch.[32]

Eyth also suggested that one of the first steam engines to reach Egypt in the early 1850s was a conventional Cornish beam engine purchased by Abbas Pasha (Viceroy 1848–54) from Kennedy and Co. of Liverpool. It was originally intended to lift water from a well in the eastern desert (present-day Abbasiya, Cairo) to one of the palaces of the Pasha, who had the proclivity of staying in secluded places. It could not lift the water to the higher planes of the palace, so it was moved to Abbas's estates in Shubra for irrigation. In the mid-1860s, Eyth saw it in operation, a 100 nominal horsepower machine, irrigating 1,600 to 1,800 acres.[33] It was an imposing figure with its long chimney, dominating the landscape outside the window of his rooms in Shubra in 1863 – that 'country's first large steam pump', to be dwarfed shortly after, he added, by the 200 and 300 hp Cornish pumps Ismail ordered that year from Europe for his Middle Egypt estates.[34]

So far, from this inventory of anecdotal observations of the 1850s, we have sugar-refining engines, mud machines, and irrigation pumps. A fourth category is the steam-powered cotton-ginning process that started to proliferate in the Delta around the late 1850s, but we will discuss this in detail in the next chapter.[35]

30 Eyth, *Das Agricultur-Maschinenwesen*, 13.

31 Ibid., 19. In the mid-1860s, larger and more ambitious steam-run dredging machines were used for digging the Suez Canal. See Chapter 7.

32 David Landes, *Bankers and Pashas: International Finance and Economic Imperialism in Egypt*, London: Heinemann, 1958, 83.

33 Eyth, *Das Agricultur-Maschinenwesen*, 20.

34 Max Eyth, *Der Kampf um die Cheops-Pyramide [The Battle for the Cheops Pyramid]*, Heidelberg: Carl Winter's Universitätsbuchhandlung, 1902, Chapter 1.

35 Roger Owen, *Cotton and the Egyptian Economy, 1820–1914*, Oxford: The Clarendon Press, 1969, 78.

As Eyth said, these early steam machines were more prototypes subject to trial and error than the product of stabilised machinery production lines. Prototypes kept showing up over the next twenty years, when Ismail imported machinery for irrigation and sugar production, among other purposes, that was in several cases new inventions whose inventors were keen to try them in Egypt and other countries. For Ismail's reign, I pursue this point further in Chapter 9.

The appearance of steam machinery in the Delta at the turn of the 1850s seems to have been contingent on other factors unconnected to a predetermined decision to modernise agriculture along a linear evolutionary trajectory of an imagined programme of industrialisation. In the case of the mud machines of the Mahmoudiya canal in 1851, it is likely that, after the end of Mehmed Ali's ultimately failed industrialisation project, these very few experimental steam machines were imported and installed to do essential work that would have otherwise required a daunting process of recruiting and deploying corvée labourers periodically. Peasants were still available for corvée, no doubt, but, by the 1850s, the transition to large landholdings – preceded by no less than a decade of large-scale *tasahub*, peasants fleeing villages en masse – and the undermining of the centralised agricultural production system of monopoly had left the irrigation infrastructure in disrepair. The cumulative effect was dire, to the extent that in 1855 the Delta land feeding on summer canals – the deeper sloping irrigation arteries that required the most maintenance – had shrunk by 300,000 feddans, compared to 600,000 feddans in 1837.[36] Meanwhile, the Mahmoudiya canal, besides its role in irrigation, was a major highway between Alexandria and Cairo, and it must have been vital to keep it in good repair – hence Abbas's decision to import several experimental dredging machines in the early stages of development as a quick fix.

The story of Halim's steam engines for dredging is probably different. In the latter case, we have a prince who, for ideological and personal reasons (explained in Chapter 7), had the intention, and the money, to invest in modern steam machinery on his private estates. But also imagine how much labour time the dredger could have freed up for the peasants toiling on his land if the machine turned out to be efficient: a

36 Ibid., 71. For a detailed account on the failure of the summer canals around mid-century, see Alleaume, 'An Industrial', 339.

mix of ideological preferences and economic practicalities for the enterprising prince. The first Cornish steam pumps? We know from slim evidence in the British press around 1846 that Abbas's grandfather Mehmed Ali had ordered in 1845 two Cornish pumps already, and Ibrahim ordered eight more in 1846 on his visit to Manchester, and these may have been the first of their type. Finally, we can also assume that the success of the new steam machinery in Middle Egypt for sugar at the turn of the 1850s gave the Pasha confidence in using more pumps in the Delta. In other words, the early steam pumps came in across 1845–50 in fits and starts. Can we think of certain motives that were stronger than others?

The cluster of motives that stands out as most relevant to explaining the early stabs at using steam power for irrigation in the Delta are those related to the cotton-production imperatives and its international market, in combination with the ecological factors of the epizootic and deteriorating summer irrigation infrastructure. This cluster is in itself complicated and multifaceted. We can trace it back to the mid-1840s large estates/cattle murrain combination leading to the power crisis. With the destruction of the country's animal power in combination with land redistribution, human labourers must have been more engaged in day-to-day agricultural operations, like powering the *saqiyas* and ploughing the land, with fewer hands to spare for maintaining the canals. Simultaneously, Mehmed Ali by 1845 had declared his intention (never realised in his reign) to revolutionise the Delta irrigation system by building the Delta barrages, after witnessing the steam pumps in action at the renovated Alexandria docks that year – the likes of which, he thought, would have made it more feasible to construct the barrages. Rich landholders, owners of *chifliks* and *'uhdas*, must have been thinking along the same lines: how to navigate the straits of failing summer irrigation and potentially recurring episodes of the murrain? The steam engine was the answer, as Ibrahim's industrial visit to England in 1846 made plain, but the Pasha was weighing down the new landholders with his in-kind and in-cash taxation demands, likely ruling out decisions on large investments like Cornish pumps of 100 hp calibre and their costs of operation, those of coal, mechanics, and direct operation labour.

By the late 1840s, however, two changes occurred: first Ibrahim and then Abbas removed all earlier restrictions on cotton sales. The Pasha's manoeuvres to circumvent free trade thus came to an end.

Second, in the early 1850s cotton prices started to recover from a slump in the 1840s – from an average of 8.75 dollars/quintal for the years 1845–9 to an average of 11 dollars/quintal across the 1850s – to the extent that some of the wealthiest landholders shifted their land from the lucrative rice to cotton.[37] Cotton prices were recovering and international trade windows of opportunity for the landholders were being opened up widely. Hence, investments in beam engines for irrigation became less of an adventure.

This period also witnessed a great divide in the quality of cotton cultivation along lines of class and landholding. The highest prices at the docks were fetched for the *Jumel* cotton – which was able to compete with high-quality American cotton, the Sea Islands variety – while the less wealthy peasants who continued to grow the crop had their production devolving to what came to be known as the *Balli* type. While the first needed frequent irrigation across the year, the latter was only irrigated twice annually. With cotton cultivation being a labour-intensive process, the latter method allowed cotton to be grown with fewer hands: less-frequent irrigation led to fewer weeds that needed to be uprooted, and fewer hands and oxen for handling the irrigation itself. The *Balli*'s labour costs and irrigation volume versus cotton price trade-off would likely have worked out for the less fortunate cultivator.

The *Jumel* cotton at the turn of the 1850s surely had a different and aggressive profitability profile: growing on large estates with inherently better irrigation access and more powerful proprietors who used their power to bring in labourers, and their money to cover higher production costs. With the inflation of the international cotton prices around the same time, more of their land was turned to cotton, hence the need for more water from deteriorating summer canals that were notorious for their excessively weak sloping during lower Nile months, and greater inclination to pile up silt.[38] The beam engine must have been the answer. And who to import it before the richest and most powerful? By the

37 Prices: Egyptian Ministry of Interior's book of statistics for the year 1873 (Cairo, 1873), 172–3. Cited in Owen, *Cotton*, 73 (Table 5). Owen, *Cotton*, 74–5; Fowler, *Report*, 8–10.

38 On the cotton price trend, *Jumel* versus *Balli*, removing Mehmed Ali's restrictions: Fowler, *Report*, 11; Owen, *Cotton*, 72–5. On summer canals disrepair as motivating investment in steam pumps: Alleaume, 'An Industrial', 339.

mid-1850s, around three eighths of all cotton produced across the Delta came from the estates of Abbas, his son Ilhami, and Ibrahim's heirs.[39] The cotton exported from Alexandria to Europe and other destinations rose sharply, from around 120,000 cantars (94 lbs in a cantar) and 257,500 cantars (1848 and 1849 respectively; higher grain prices at the time might have also motivated a transient shift away from cotton), to an average of 370,000 cantars (for 1850 and 1851) then 670,100 cantars (in 1852).[40] After that, the volume of cotton hovered around half a million cantars throughout the 1850s.[41]

If Abbas, his son Ilhami, and Ibrahim's heirs – including the younger Ismail, beside Ahmed Pasha and Mustapha Pasha – produced three eighths of all cotton across those years, one should expect that considerable parcels of land started to get their water from beam engines. Overall, Ghislaine Alleaume's conclusions on the early proliferation of steam technology in the 1850s stand as credible, even if we keep in the picture the sugar and sugar cane steaming question around the same time in Middle Egypt: 'the large estates were the first agricultural sector to mechanise', and 'the industrialisation of [agricultural] production began at the pumping station.'[42]

Conclusion

The sugar factories in Middle Egypt, accompanied by the earlier generation of steam irrigation pumps (Cornish/beam variety), were likely the bearers of the earliest substantial experimentation in steam machinery in Egypt after Mehmed Ali's hybrid industrialisation from the 1820s and 1830s. The opportunities of global trade in refined sugar pushed Abbas Pasha and others in the Viceroyal family to introduce steam machinery

39 Owen, *Cotton*, 75.

40 It is noteworthy that rising grain prices after the repeal of Corn Laws in England in 1847 might have also partly led to the slump in cotton exports in 1848–9, as Egyptian cultivators invested more in growing the less labour-intensive grains for Europe. See Fowler, *Report*, 10, 17. The same grain exports profitability trend was maintained again in the mid-1850s with the Crimean War, but did not seem to have an impact on cotton productivity in Egypt, which was maintained at around half a million cantars across the decade.

41 Ibid., 10–11.

42 Alleaume, 'An Industrial', 338–9.

in sectors that needed the technology for them to be able to sell the commodities these sectors produced: only white refined sugar in purist form – achieved through steam-powered processes – was in great demand everywhere. The same process was witnessed simultaneously or slightly later, likely on a bigger scale, across the cotton-growing estates of the Delta. With the end of the global cotton prices slump of the late 1840s and the imperatives of the *Jumel* cotton that required intensive irrigation on the lands of the newly formed landed-nobility class, the beam engine started to show up to bring more water from deteriorating summer canals.

But the *Jumel* and *Balli* cottons needed more processing besides irrigation before they reached the docks. The ginning of cotton led to more substantial proliferation of steam machinery in the second half of the 1850s and the early 1860s. The steam machinery for ginning cotton led to further expansion in its cultivation, especially with the cotton boom associated with the American Civil War in 1861, under conditions of labour shortage relative to the labour-intensive cash crop. This expanding and progressively more mechanised/steamed cotton economy relied greatly on the freeing of trade, tax monetisation, and the proliferation of European fossil capital in Egypt under Sa'ed Pasha's watch (1854–63). We turn to this next.

4

The 'Want of Capital'... The Cotton Gin (1855–63)

By the start of the 1850s, Abbas Pasha was in the castle. The grandchild of Mehmed Ali stands accused by most historians of having been an ultra-conservative, closed-minded man who brought a few years of darkness and ignorance after decades of marching towards civilisation and grandeur. He closed schools, dismissed European experts in the administration whom Mehmed Ali had brought in for their services, and was generally against progress.[1] Nubar Pasha in his memoirs said that his name (Abbas, which carries connotations of bleakness and sullen faces) hovered heavy on the Nile valley, and when Sa'ed (meaning 'happy') came to power in 1854, the peasants had their fair share of the latter's name.[2]

Abbas might have been a xenophobic, suspicious, and cruel man who saw conspirators lurking in all shadows, but the procedures he took towards shutting down schools and dismissing foreigners were trends that already had been set in motion by the latter years of his grandfather's reign. The schools Mehmed Ali opened were not centres for

1 For example, Afaf Lutfi Marsot, *Egypt in the Reign of Muhammad Ali*, Cambridge: Cambridge University Press, 1984. Most extreme anecdotes and views on Abbas's character and his bleak age were waged by French men and women in Egypt. See for example: Nassau William Senior, *Conversations and Journals in Egypt and Malta, Vol. 1*, London: Sampson Low, Martson, Searle & Rivington, 1882.

2 Nubar Pasha, *Muzakirat Nubar Pasha* [Arabic translation by Jarou Robert Tabaqian of: Mémoires de Nubar Pacha], Cairo: Al-Shorouq, [1983] 2009.

teaching and civilising the population for the sake of the very idea, but more like factories to hatch military personnel, technicians, and administrators of the army, workshops, and agricultural estates.[3] The grand Pasha himself closed down many of the schools during the power crisis years of the early 1840s. The European experts Abbas dismissed were mostly French military experts and engineers who became redundant with the radical downsizing of Ali's army and the failure of his industrialisation project. The Europeans who tried to infiltrate the interior during his reign were mostly traders and representatives of commercial houses who wanted further access to production centres, following the general trend that started in the mid-1840s with the coercive freeing of trade by the British Empire and the end of monopoly.[4]

In short, let us allow for more continuities between Mehmed Ali's latter years and Abbas's short reign. Along these lines, Abbas's policies on agriculture and infrastructure can make more sense. In 1849, he re-confiscated many villages that had been held as *'uhdas* since the mid-1840s, as their noble holders kept failing to deliver taxes in arrears. Between one seventh and three quarters of *'uhda* lands and villages were thus treated that year.[5] Perhaps the gap between these two estimates is due to what he did with the confiscated lands: reassigning some of them to village heads by force, so that more peasants, once again, were labouring on village lands. On the other hand, some of the *'uhdas* confiscated were reassigned to their holders in the form of *rizqa* land, that is, granted to them as their own estates, with usufruct rights for life, entailing that many peasants on these *'uhdas* were transferred to *rizqa* as sharecroppers and wage labourers.[6] Meanwhile, expansion in agrarian infrastructure was halted, with some exceptions. The Delta

3 See generally: Khaled Fahmy, *All the Pasha's Men: Mehmed Ali, His Army and the Making of Modern Egypt*, Cairo: AUC Press, 2002; Timothy Mitchell, *Colonising Egypt*, Oakland, CA: University of California Press, 1991.

4 See generally on this, and on closing schools and dismissing foreigners in administration, David Landes, *Bankers and Pashas: International Finance and Economic Imperialism in Egypt*, London: Heinemann, 1958, 81–2.

5 One seventh according to Nubar, two thirds–three quarters according to the British consul, in Kenneth Cuno, *The Pasha's Peasants: Land, Society, and Economy in Lower Egypt (1740–1858)*, Cambridge: Cambridge University Press, 1993, 160.

6 On reassignment of *'uhda* as *rizqa* to the same proprietors: Amin Sami, *Taqwim Al-Nil 3.1 [Survey of the Nile Vol. 3.1]*, Cairo: Matba'at Dar Al Kutub, 1936, 33; Cuno, *The Pasha's*, 160.

barrages project practically stopped, and irrigation infrastructure continued to deteriorate.[7]

With these continuities, ultimately two political motives loomed large over his years: strained relations and power dynamics with the Sublime Porte and European powers, against which he reacted with a wish to entrench his power and support it with greater wealth for himself and his son Ilhami, whom he hoped would be his successor. Mehmed Ali's experts were mostly French men, and Abbas wanted to curry England's favour against perceived or real French-Ottoman conspiracies. The greatest infrastructural project he started, the Cairo–Alexandria Railway, was at the heart of this. During a meeting with a British consular officer in 1850, the Pasha reportedly expressed his intention to build the railway. He said that

> he was not going to make railways for his own accommodation, or because the Egyptian Treasury has any surplus fund to employ in such operations; but because [the British government] expressed a wish to that effect, and for the purpose of improving the overland communication, and so facilitating the intercourse between England and its Indian possessions.[8]

Another British consular officer sent a letter to the foreign secretary, Lord Palmerston, requesting support to Abbas Pasha against 'political intrigues' from the French, while also stating that a railway between Cairo and Alexandria was 'very beneficial to our Indian communication and to Egypt itself'.[9] In the same vein, amid the early starts of the railway – finished after his death, in 1856 – he might have tried to relaunch the great monopolies to consolidate his wealth and power, perhaps thinking he had already appeased the British with the railway project and would thus gain their support.[10] In 1854, a few months before his

7 Landes, *Bankers*, 82.

8 FO 474/7A, 3. In September 1851, Abbas issued an order for starting the works for the railway. Sami, *Taqwim, Vol. 3.1*, 42.

9 FO 474/7A, 1.

10 Cuno presented archival evidence that suggests that Abbas's monopoly manoeuvres were overstated by European commercial houses, traders, and diplomats, and that he might have merely wanted to profit for a while from the grain famine induced by the Crimean War. Consular correspondences cited in Cuno, *The Pasha's*, 183.

assassination, he issued an order that banned all sales of land produce except for government officials, which was never fully applied, and introduced a 10 per cent tax (*'ushr*) on *chifliks* and other lands of the nobility. He was perhaps trying to create a new monopoly to support him against internal and external conspirators, by means of collecting more tax, increasing profitability from agricultural production, and ultimately consolidating Ilhami's position as successor later on.[11]

These were the conditions under which steam engines started to appear on the Nile in the early 1850s – as power appliances to irrigate and process sugar cane and cotton for wealthy men, including the Pasha and his son, on their extensive private plantations. The description of Abbas as selfish and reactionary in his policies is thus given more context.[12] The trend of using steam machinery for consolidating personal wealth, of the Pasha in power and the most powerful men around him, continued in the reign of Sa'ed, but now in combination with greater penetration of trade and money.

After Abbas was assassinated by his servants in the bathtub of one of his secluded palaces in 1854, Sa'ed immediately reversed the departing Pasha's policies. Sa'ed was said to be the most Westernised nobleman in the country thus far. He had a partial European education and could speak English and French. He surrounded himself with European friends and arguably believed in the spirit of the age: free trade.[13] However, it could have been the case that he was motivated also by the mounting lobbying by trade-freeing agents from Europe. Globally, the turn of the 1850s was the stage of a major economic crisis in Europe. With it, idle capital in Europe was recirculated as investments and loans to a wider global circle, with many potential sites newly seen as nodes for investment returns higher than the conventional ones well worn thus far.[14]

Trade was indeed liberated more than ever across the first half of the century under Sa'ed's watch. He struck down the various small customs levied on traders in the interior. More importantly, in tandem with the

11 On Abbas's monopoly: Roger Owen, *Cotton and the Egyptian Economy, 1820–1914*, Oxford: The Clarendon Press, 1969, 67–8. '*Ushr*/tenth tax introduced in 1854: ibid., 68.

12 Landes, *Bankers*, 82.

13 For example, ibid., 82–4; Cuno, *The Pasha's*, 184.

14 Aaron Jakes and Ahmed Shokr, 'Capitalism in Egypt, Not Egyptian Capitalism', in *A Critical Political Economy of the Middle East and North Africa*, edited by Joel Beinin, Bassam Haddad, and Sherene Seikaly, Stanford: Stanford University Press, 2021, 127.

economic crisis of western Europe and the penetration of its financial capital on a global scale, he ordered the payment of all tax in coin, not in kind. This radical economic procedure was also likely contingent on the concession Sa'ed granted to Ferdinand de Lesseps to excavate the Isthmus of Suez, for which an international joint-stock company – La Compagnie Universelle du Canal de Suez – was formed, and in which Sa'ed bought over 40 per cent of the shares. So, after decades during which the Pasha's granaries and storehouses contained monopoly-collected produce, these buildings were shut down. But not overnight in Upper and Middle Egypt, where taxes in kind were revived shortly after, given greater scarcity of money in the south, and an extensive insurrection in Middle Egypt by Bedouin groups, which might have hindered the shift to cash taxation.[15] However, the full shift to monetised tax was permanent in the Delta, the region that at the time had over two thirds of all productive land and the majority of the population.[16] This meant greater penetration of market relations in the interior, evident in the return of moneylenders and traders who advanced capital and seed to less wealthy growers, a process that Mehmed Ali stopped for decades. It also means, as Cuno argued, that money lending and growing crops on credits did not originate in Sa'ed's reign, but was a return to an old process, this time coming mainly from Alexandria, the city that turned in his age into 'an outpost of European capital'.[17]

Meanwhile, landholding laws (1847, 1855, and 1858) accompanied the trend of increasing liberation of trade with more and more security

15 Cuno, *The Pasha's*, 184. Cuno did not account for further complications regarding taxation in the south in Sa'ed's reign, including a vicious and seemingly extensive war with Bedouin groups across Middle Egypt, Fayoum to Assiout, where army divisions equipped with modern cannons were sent south, one of which was headed at some point by Sa'ed himself. He was close to losing his life at one of the clashes. A reason for this conflict was reportedly that Sa'ed imposed regular peasant taxes on Bedouin tribes instead of the 10 per cent regularly associated with large estates (Mehmed Ali imposed 10 per cent tax only on lands held by powerful Bedouin tribes, and pardoned them from corvée duties, to gain their willing support in his wars). This complicates the picture of Sa'ed's reign concerning taxation and resistance to it in the south, but it remains outside the scope of this book. For a detailed study based on rich primary material (Viceroyal directives and military correspondences), see Ibrahim A. Almorsi, 'Harakat Omar Almasry fi 'Ahd Sa'ed [The Omar Almasri Insurrection in the Age of Sa'ed]', *Journal of the Faculty of Humanities, Mansoura University* 27, 2000, 371–508.

16 Cuno, *The Pasha's*, 184.

17 Ibid.

granted to landholders, officialising their hold and further securing inheritance.[18] With the liberation of trade and greater penetration of European capital, steam engines kept coming, for Sa'ed and the wealthiest in the country, to the extent that his reign was described in contemporary and later sources as the years of the early application of steam irrigation, model modern farms, and expansion in steam-powered cotton ginning.[19] This process of monetisation and Sa'ed's embrace of free trade shaped how steam machinery kept being bought on credit from European banks and capitalists, increasingly tying Egypt during the third quarter of the nineteenth century to the twists and turns of global trade in commodities and financial capital. The cotton gins at the turn of the 1860s were the driver and embodiment of how financial and fossil capital came together to change the lay of the land.

Pre-boom Gins

John Ninet, a Swiss agricultural expert and big landholder in Egypt whom Mehmed Ali invited in 1839 'to help advance agriculture', later nicknamed the Swiss farmer, reported seeing the American McCarthy steam-powered cotton gins in Egypt and England in 1854–5.[20] In 1860, the Delta possessed the capacity to process 12.5 per cent of its cotton through the McCarthy gins, according to estimates from Thomas Fowler, an Alexandria-based

18 See ibid., 193–7 for a more nuanced discussion on the meaning and implications of mid-century landholding laws. For a critical review of landownership laws, their relevant archives, and how historians made sense of them, see Mohammad Hakim, 'A Multiplicity of Rights: Rural-Urban Contradictions in Early Nineteenth-Century Egyptian Land Ownership', in *Money, Land and Trade: An Economic History of the Muslim Mediterranean*, edited by Nelly Hanna, London: I. B. Tauris, 2002. Another grounded discussion on landholding and property laws, and their place in the power dynamics of the progress of landholding and landownership in Egypt across the nineteenth century: Timothy Mitchell, *Rule of Experts: Egypt, Techno-politics, Modernity*, Oakland, CA: University of California Press, 2002, Chapter 2.

19 Landes, *Bankers*, 84; Max Eyth, *Lebendige Kräfte: 7 Vorträge aus dem Gebiete der Technik* [Living Forces: 7 Lectures from the Field of Technology], Berlin: Julius Springer, [1905] 1908, 214, 218.

20 Cherine Chehata, 'John Ninet et le Militantisme Etranger', *Journal of Humanitarian and Literary Sciences, Ain Shams University* 27:3, 2021, 51–70. John Ninet, 'La Culture du Coton en Egypte et les Filateurs Anglais', *Revue des Deux Mondes* 12, 1875, 577–96, 583.

English merchant who surveyed the Delta in 1859–60, months before the American Civil War and the great cotton boom.[21] His survey was in preparation of a report for the Manchester Cotton Supply Association.[22] With the steady march of the liberation of trade, monetisation, and penetration of foreign capital in the Delta, a mix of wealthy landholders and European capitalists established ginning factories. It started, however, with European capitalists before 1861. Messrs Kruipser and Co. from Germany had by 1859 or 1860 a mill of 100 roller gins powered by steam engines, at Abusir in Gharbiya. Three other smaller steam-powered facilities were operating simultaneously: one at Mansoura, fifty cotton gins, owned by Messrs Ralli and Co.; a mill of forty gins at Kafr Al-Zayat, owned by Mr Pacho; and one in Mahalla Kubra, composed of thirty gins, owned by Ilhami Pasha, Abbas's son. Two other ginning mills were in the course of erection in 1860 or earlier, Fowler observed, one at Sammanoud, by a certain Bedrawich Bey (forty to fifty gins), and another in Banha, by Ilhami.[23] Overall, Fowler reported the operation of 300 steam-powered gins across the Delta by 1859 or 1860.

However, it is misleading to attribute the established and planned power gins in Mahalla Kubra and Banha to Prince Ilhami, being classified as a native wealthy landholder. For, after the death of his father in 1854, the adolescent prince instantly turned the administration of his inherited 80 million francs' wealth and extensive lands to 'experienced businessmen'.[24] By 1856, he was described by his physician as 'an old man at twenty-one'. That year, with mounting debts, Ilhami turned his lands to the Prussia-backed bank of Oppenheim, Chabert and Co. in Egypt, to manage it for twelve years. Besides the annual fees for administering this wealth, the bank might have made lucrative profits from commissions for purchases of machinery and agricultural equipment. Here was international financial capital expanding through investments in fossil-fuelled technology. Soon after, partly under impact of the bank loans Ilhami took out, and partly by the cumulative leakage from the

21 Owen seems to have dated the conduct of Fowler's survey as early as 1859. See Owen, *Cotton*, 78. As such, Fowler's survey predated the American Civil War and its associated cotton boom.

22 Thomas Fowler, *Report on the Cultivation of Cotton in Egypt*, Manchester: J. J. Sales Printers, 1861, 22.

23 Ibid.

24 Landes, *Bankers*, 112–14.

inherited wealth in the form of annual fees and commissions for the bought machinery and other modern enhancements, Oppenheim, in turn, borrowed from the Bank of Egypt. The latter had the backing of the British consulate and was notorious for having its clients 'generally finishing by being compelled to abandon their property to the Bank'.[25] With Ilhami's departure from Egypt in 1860, then his premature death the same year, the Oppenheims and the Bank of Egypt – backed by the Prussian and British consulates – pushed Sa'ed to settle the debts and protect both banks.

The Prince's assets were liquidated, and the state treasury covered the remaining amounts demanded by the creditors. In 1862, the first major state loan in Egypt's modern history was issued: 60 million francs at 11 per cent interest, the annual payment of which was tied to the country's land tax.[26] Nothing in this is to suggest that the steam-powered cotton mills of Ilhami were exclusively behind state debts and the proliferation of credit on any major scale. Instead, this affair can be seen merely as a precursor for a pattern followed over the age of Ismail: the financing of imported steam technology by state loans.

Some other cotton-growing micro-scale dynamics of monetisation–loans–steam gins were more substantial, even before the great cotton boom of the early 1860s. With the full monetisation of land tax in the Delta that Sa'ed decreed in the mid-1850s, the less wealthy and powerful *Balli* cotton cultivators, like producers of all other crops, had to pay taxes levied quarterly, in advance, regardless of their liquidity situation, that was doubtless more secure only around the times of harvest.[27]

Although Sa'ed issued several orders asking tax officials to make collections 'around times of liquidity' for the peasants, it is doubtful that this was effective in pushing back small and middle farmers' indebtedness.[28] To cover the cash tax demands, the less fortunate farmers turned to Greek and Levantine moneylenders, who levied 3–4 per cent monthly

25 All citations: ibid; on Bank of Egypt Landes cited: Foreign Office correspondences FO 142-24, no. 38, 14 July 1860.

26 James Whidden, *Egypt: British Colony, Imperial Capital*, Manchester: Manchester University Press, 2017, 33.

27 Fowler, *Report*, 22.

28 See, for example, order issued by Sa'ed to the governor of Gharbiya on 8 Muharam 1272 H [19 September 1855], in Sami, *Taqwim Vol. 3.1*, 131. In his order to the governor of Qina and Esna a year earlier, he specifically mentioned the necessity of waiting for crops to be sold before taking the quarterly taxes from peasants. Ibid., 95.

interests. As such, any delays in cotton sales meant more interest accumulating on these micro-loans. Meanwhile, manual ginning was time-consuming: one cantar was processed in a week (an acre of *Balli* produced on average two cantars), but the same task took a McCarthy roller gin only ten hours.[29]

The effect of the slower manual ginning, as Fowler gathered, was that even the most modest quantity of cotton grown by a peasant stayed whole with their seeds, heaped in the barns for months, deteriorating in quality. The farmer was thus deprived of realising the profit of his crop long after the harvest, accumulating debts in the form of more monthly interest.[30]

A peasant growing cotton, hard-pressed to sell the crop rapidly to avoid higher monthly interests, would no doubt have found in the McCarthy gins the answer to his prayers.[31] These steam-powered gins were also vital for extending *Balli* cotton to more plots of land with the fix for loan interests it offered by compressing time between harvest and sale. With cotton paying better than grains, farmers shifted from the latter to the former only when they found nearby steam-powered cotton mills that processed cotton quickly and made the balance between lower grain profit and higher cotton price margin – that was lost to monthly loan interest – tilt in favour of cotton. The director of the Prussian-owned 100 McCarthy gins in Abusir told merchant Fowler that cotton cultivators in the area rushed to make arrangements with him for ginning the produce of larger plots on their farms that they planned to dedicate to cotton, even before they started growing the crop. If they were told that the facility would accommodate their higher cotton yield in the coming season, they proceeded with the planned expansion.[32]

In short, for the *Balli* cotton grower, monetisation of the land tax brought Greek and Levantine moneylenders, and the latter brought, in their wake, European fossil capital – as well as a few Egyptian fossil capitalists – who intervened in the labour of cotton growing and ginning to compress processing time and ultimately take profits in the form of monthly interest off the hands of the usurers, on one hand, and profit from the indebted cotton producers on the other hand. The

29 Fowler reported that the 100 gins at Abusir processed 100 cantars per day; see Fowler, *Report*, 31.

30 Ibid., 21–2.

31 Evidence from ibid., 22, also concurred with in Owen, *Cotton*, 78.

32 Fowler, *Report*, 22.

steam-powered cotton gin owners in the Delta were making net profits of 20 to 25 per cent in 1860. Fowler called on the capitalists of Manchester and beyond to come forward purely as fossil capitalists to burn coal for compressing cotton-processing time, a fossil-fed industrial process that we can see now was so profitable because it worked against the sheer force of monetised taxation and exploitative dynamics of micro-loans. 'Here then, is a perfectly safe and extensive and sufficiently profitable field for investment in this country' that, at the same time, merchant Fowler argued, was also going to propel the expansion in cotton production much further in the near future, and increase Egyptian supplies to the English textile factories.[33]

Regarding the superior *Jumel* growing at the time on the larger estates, the monetisation/steam gins dynamic seen here with *Balli* was not repeated. Something different must have been happening on the big cotton plantations before and with the start of the American war. An unspecified reporter to the *Belfast Mercury* from the summer of 1861 said that the McCarthy ginning mills were only used to a 'small extent' on the 'estates of large landholders' in Egypt. If some of the princes, Pashas, and Beys that year followed the European cotton mill capitalists in the Delta and established more steam-powered mills to make profit with the peasants' *Balli* cotton, then that is another story. Perhaps some of them did but did not necessarily use it on their own produce, because 'the greater number of the Manchester spinners object to it [i.e. McCarthy-ginned cotton imported from Egypt]'. At the Alexandria docks that year, 'there are few who actually give it the preference over cotton cleaned by the old method'. The McCarthy-processed cotton 'has to be disposed at lower prices', and made for 'much slower sale' in England.[34] One can deduce from this information that the McCarthy-ginned cotton at Manchester was mostly of the shorter-fabric *Balli* type, which, even after enhancement by steam-powered uniform separation of seed and cleaning, was of lower quality than the *Jumel*. The latter, at least until 1861, must have been the 'cotton cleaned by the old method'. Manchester spinners might have changed their take on McCarthy-ginned Egyptian cotton a few years later when *Jumel* started passing through the McCarthy rollers.

33 Ibid.

34 'Egyptian Cotton', *Belfast Mercury*, 14 August 1861.

Now, why did large landholders not shift to the faster time-compressing steam-powered gins? It can be hard to get to a definite answer, but we have several probable factors. First, they were not in a rush. The owners of *chifliks* and other privileged lands of the nobility had to pay only 10 per cent tax, and, even if they were in need of loans after Sa'ed opened up fully to trade, their loans must have induced less interest compared to the 3 to 4 per cent monthly rate accepted by the smaller farmers. They had their own agencies in Alexandria that took care of the sales, and, more importantly, they took their time sorting and organising the product into several batches based on quality, a process that must have consumed considerable time in any case.[35] For example, in 1859–60, an English cotton expert was hired by the Prussian bank to grade cotton on Ilhami's estates. As soon as the labourers returned with their day's picking from the field, he started his work, sorting cotton in different piles according to colour and ripeness, each fetching its own price.[36]

Second, as we have seen, the large landholders were more successful, through using their power, in securing 'hands' on their fields even under conditions of labour scarcity. Their labourers were not subject to state corvée, and by the late 1850s, the new institution of the *'ezba* was secured, tying peasants working on large estates even more tightly to the lands of the rich. The *'ezba* originated as an enclosure of houses built by the Pashas for labourers who were relocated with their families to live in the middle of their extensive plantations, to work their land and grow food on small subsistence plots that they were granted.[37] Many or most of these *'ezba* peasants must have been tax-, conscription-, and corvée-fleeing peasants who, over the earlier decades, turned from landed

35 Owen, *Cotton*, 77.

36 Ibid., 75. Decades of experimentation with various cotton strains, Owen maintains, polluted the *Jumel* fields with other types. It was common to see white, yellow, and brown cotton growing together on the same field.

37 For a more detailed discussion on the *'ezba*, see Alan Richards, 'Primitive Accumulation in Egypt, 1798–1882', *Review (Fernand Braudel Center)* 1:2, 1977, 3–49; Mitchell, *Colonising*, Chapter 2. Richards's categorisation of *'ezbas* as originating in the 1850s, however, might be wrong. There is evidence that the *'ezba* system goes back at least to the 1830s. The word shows up occasionally across Mehmed Ali's published correspondences as a taken-for-granted unit of agrarian production. In the early 1830s the Pasha allowed the establishment of *'ezbas* in Sharqiya, which was suffering the fleeing of peasants in great numbers to Syria and Palestine. See Order to Sharqiya police director, 21 Ramadan 1245 H [15 March 1830].

producers to hired hands. So, in this respect, power over the peasants made the steam power of the mechanical gin redundant in the cotton-production process on large estates.

Finally, an identifying feature of *Jumel* was its frequent irrigation. By the 1850s, summer irrigation was in a worse condition than it had been since the 1820s. Landes estimated that 'Sa'ed's relatives' were the agricultural producers who championed the progress and increase in crops during the 1850s, against the odds of deteriorating irrigation infrastructure, including the problematic summer canals that were needed for the perennial cotton crop.[38] One can only assume that those rich landholders, since they were not responsible for any irrigation infrastructure, and not encumbered with a sizable land tax, were catering to their *Jumel* by importing the Cornish beam engines to irrigate the cotton. As such, they must have made more than enough investments in irrigation pumps alone, or, more probably, their cotton-planted fields reached a most satisfactory profitability balance with available 'hands' picking, sorting, ginning, and pressing their cotton.

The other potential motives that we can glean from possible combinations between any of these three sets of factors for the *Jumel*-producing rich proprietors keeping away from the McCarthy steam-powered mills can make for a long list. Suffice it here to assume that, after this brief period of *Jumel* manual ginning, the American war must have changed that balance, leading to more steam-powered cotton mills on the lands of the rich. The craze for cotton on the international market pushed the prices to unthinkable peaks, and other solutions needed to be found to grow and sell more cotton.

The Cotton Boom... 'The Want of Capital'

Over in America, in April 1861, war broke out between the Confederacy and the Union. First, the Confederacy banned cotton exports to elicit British diplomatic recognition, then, by the time this policy proved useless, the Union had enforced a blockade that prevented most cotton exports from the South. By 1862, no American cotton reached British manufactures – that sustained at least 20 per cent of England's

38 Landes, *Bankers*, 84.

population – or other cotton-spinning centres across Europe, except for the occasional small dispatches by smugglers.[39] With the great slave-based cotton production of America disrupted, the decades-long global economy of cotton and textiles fell into disarray. By early 1862, with total imports to England falling by 50 per cent (and 96 per cent from America), earlier stocks were consumed, sold, and resold at relentlessly higher prices pushed by feverish speculation. By early 1863, around 25 per cent of the population of Lancashire (over half a million) was unemployed. The cotton famine was accompanied by the rioting of unemployed workers across several British cotton towns. Local authorities sent requests to the home secretary asking how they could get military assistance if things got out of control. The military was indeed deployed. In France, American imports fell by 94 per cent between 1860 and 1863. That year in Normandy, 60 per cent of the looms were idle. A national-level committee estimated that over 250,000 textile workers lost employment. In the Alsace cotton towns, posters clandestinely put up on walls declared 'Bread or death'. Across Germany and Russia, and other lesser textile centres, the same pattern held. In Saxony alone, 100,000 workers – one third of the cotton-spinning workforce – became unemployed. In Russia, around 75 per cent of all cotton-spinning mills closed down in 1863. Alongside these dramatic changes, and although global cotton circulation was reduced by roughly 50 per cent, owing to the almost total absence of American cotton, its overall monetary volume rose from around £40 million to £84 million. Everyone with any amount of money to spare was speculating in cotton. 'Doctors, parsons, lawyers, wives and widows' in England and across the continent became cotton traders.[40]

Across the globe, a race took off to grow cotton to feed the cotton famine. Brazil, Ottoman lands, Argentina, West Africa, Australia, West Indies, among other places, turned to cotton.[41] But India and Egypt stood out, by far, in terms of the long-lasting impacts on land conversion and cotton mono-cropping. At least quantitatively, the statistics on

39 Corey Ross, *Ecology and Power in the Age of Empire: Europe and the Transformation of the Tropical World*, Oxford: Oxford University Press, 2019, 29. Sven Beckert, *Empire of Cotton: A Global History*, New York: Vintage Books, 2014, 246.

40 Beckert, *Empire of Cotton*, 249. All information in paragraph: 245–9.

41 William O. Henderson, *The Lancashire Cotton Famine, 1861–1865*, Manchester: Manchester University Press, 1934, Chapter 3.

transformation to cotton growing in India with the American Civil War greatly resemble the situation in Egypt. Indian cotton was, however, lower in quality, its market highly volatile, and more demanding in terms of transportation and local purchase across the subcontinent.[42]

In Egypt, the impact of the American war was almost instantaneous. After the steady half a million cantars of cotton of the 1850s, production rose to over 800,000 cantars in 1862, then sailed across the 1 million threshold to 1.29 million, 1.74 million, 2.5 million, and 1.78 million cantars across the years 1863–6.[43] Land on which cotton grew was expanded fivefold from a quarter-million feddans to at least 1.25 million feddans, over the same years, whereas all Nile Delta lands were around 3 million feddans.[44] By the cotton sale season of 1862, the price fetched at the Alexandria docks was treble or quadruple the price before the boom, hovering around 50 dollars per cantar.[45] It was said that two of Ibrahim Pasha's heirs, Ismail and Mustapha Pashas, became 'the two largest producers [of cotton] in the country' that season,[46] then Ismail was said in 1863 to be the biggest cotton producer in the country, with 40,000 cantars.[47]

As early as 1861, everyone in Egypt was anticipating the coming cotton price inflation. The American consul had the impression, that early, that the poorest of peasants were fully aware of the American war-induced global cotton crisis and would not let go of their cotton, if bought on their lands, at prices considerably lower than the rates at the Alexandria docks.[48] Cultivators, big and small, were extending cotton more and more on their lands. Perhaps the new ginning factories of the Delta helped sustain that expansion of the *Balli* cotton at least. By late 1863, big zones of the Delta were turned entirely to cotton. For example, in Sharqiya, the population of the capital city of Zaqaziq had reportedly

42 For a comparative narrative for both countries in terms of cotton, capital, and ecological impacts into the twentieth century see Ross, *Ecology*, 34–44. On cotton production in India in the 1860s: Peter Harnetty, 'Cotton Exports and Indian Agriculture, 1861–1870', *The Economic History Review* 24:3, 1971, 414–29.

43 Owen, *Cotton*, 90, Table 7.

44 Ibid., 89.

45 'The Egyptian Cotton Crop', *Paisley Herald and Renfrewshire Advertiser*, 4 October 1862.

46 Ibid.

47 'The Cotton Trade of Egypt', *Liverpool Daily Post*, 11 June 1863.

48 Cited in Owen, *Cotton*, 94–5.

quintupled, from 4,000 inhabitants to 20,000, between 1859 and 1863. Cotton was the reason: 'Barges, camels, and donkeys, laden with cotton, may be seen in all directions.'[49] Zaqaziq turned to a huge Delta cotton trade centre, one of several boomtowns:

> The Arabs convey it to Zagazig for sale, where the railway station is completely encumbered with it [. . .] the town [became] a great cotton mart, [and] has acquired considerable importance [. . .] Almost all the Bedouins and farmers have abandoned their former cultivation and devoted themselves to that of cotton.[50]

Steam-powered gins followed suit. By late 1861, the secretary of the Manchester Cotton Supply Association (more from him shortly) estimated, upon his inspection of the Delta, that the total steam-ginning capacity had doubled since Fowler's report of 1859–60, to 600 to 700 roller gins, which were able to process 25 per cent of the total Egyptian cotton crop.[51] Two years later (1863), the total steam-powered processing capacity applied to one third of the crop, even though the crop itself had more than doubled by then. Steam-ginning investments were reportedly so great by 1863 that Samuel Smith, the British cotton broker who came from Liverpool and toured 'interior cotton markets' of the Delta that year, forecasted that, by 1864, 'the bulk of the crop will be cleaned by them [steam-powered gins] the next year'.[52] He even estimated that steam-powered ginning factories would soon net substantially lower profits, since he saw several new factories being set up that season, many of them in the 50–100 gins range.

British capital and expertise invested in steam-powered ginning increased in tandem. For example, in late 1863, engineer Kusel of Liverpool headed to Egypt for the first time to work at a large cotton-ginning factory in Sharqiya owned by three partners. One of them was

49 'Cotton in Egypt', *Dublin Evening Mail*, 29 December 1863.

50 Ibid.

51 'Cotton Culture in Egypt', *Public Ledger and Daily Advertiser*, 3 October 1861.

52 'The Cotton Trade of Egypt', *Liverpool Daily Post*, 11 June 1863. On Samuel Smith, the middle-class cotton broker who later became a Liberal Member of Parliament, see Donna Loftus, 'Time, History and the Making of the Industrial Middle Class: The Story of Samuel Smith', *Social History* 42:1, 2017, 29–51. 'The Cotton Trade of Egypt', *Liverpool Daily Post*, 11 June 1863.

an Egyptian physician married to a British woman; the other two were British citizens. Later, in Sharqiya, he met and made friends with two fellow British engineers who were hired by a cotton-ginning factory in the same district.[53] The American traveller N. Matson noted in 1868 that 'within the last few years a number of English capitalists have been engaged in the culture of cotton and in many of the towns along the Nile, the smoke can be seen rising from their steam cotton gins'.[54]

But while steam-ginning enabled the processing of expanding cotton fields, the growing of cotton on those fields brought more steam power: the Cornish pumps. Samuel Smith, the cotton broker, in his survey of 1863 published in the *Liverpool Post*, did not stop at accounting for the expansion in mechanised cotton ginning. His detailed analysis of the obstacles to more expansion in cotton, far beyond 1 million cantars per year, hinged on irrigation power. The 'irrigational power' of Egypt was everything, in terms of cotton quantity and quality.[55] For quality, 'where irrigation is scanty the staple is short and dry, while an abundant supply of water makes it long and silky. The amount of cotton that can be grown in this country depends, therefore, upon the facilities of irrigation far more than upon any other circumstance.' In terms of quantity, the cotton crop on hundreds of acres, Smith noticed, was successfully and generously irrigated by any single steam pump, compared to using so many *saqiyas* on the same extent of land, each of which brought to the land a 'mere driblet' with the 'two bullocks' powering it. The Pashas – 'who own large tracts of country – are now getting them [steam pumps] out from England'.[56] He was told that, on any given day, 100 such high-horse-power pumps, of the kind that irrigated hundreds of acres, were operating across the fields of the Delta.

Smith noted that the richest landholders were more able to purchase steam pumps for structural reasons. Contrary to the regular peasant, with small and medium landholding – up to 600 acres in his categorisation – that accrued heavy taxes to the state, and indebtedness, the 'fee-simple Pashas, who are, generally, members of the reigning family', held

53 Kusel (Bey), *An Englishman's Recollections of Egypt (1863 to 1887)*, London: John Lane, the Bodley Head, 1915, 5, 37.

54 N. Matson, *Beyond the Atlantic, or Eleven Months Tour in Europe, Egypt and Palestine*, Illinois: Republican Job Printing Establishment, 1870, 209.

55 'The Cotton Trade of Egypt', *Liverpool Daily Post*, 11 June 1863.

56 Ibid.

lands that were 'received [. . .] in grants from the Government'. As such, 'most of these Pashas are large cotton growers, and they are the chief importers of pumping machines'.[57]

However, Smith added, a radical overhaul of the cotton Delta scheme was needed with much more capital:

> Till capital is liberally applied to the country for opening up new [irrigation] canals and supplying pumping machines, the increase of cotton cultivation cannot go much further, for already it has exhausted the irrigational power of the country.[58]

Concerning labour, the need for 'hands' on cotton fields was so great that he blamed the French-driven Suez Canal digging project that engaged 20,000 peasants in corvée any given day, for a serious drain away from expanding cotton production. For cotton capacity to expand beyond the limits, Egypt needed 'a stable and enlightened Government which encouraged the influx of population and capital'. Only with this, the Delta 'might and would become one great cotton field'.[59] In short, Samuel Smith the cotton broker was echoing what several prominent cotton experts had been saying since the start of the war about how Egypt could make up for the lost American market. We turn to some of them next.

With the war, while producers, traders, and speculators in cotton made fortunes, manufacturers frantically looked for other non-American lands, without a slave-powered plantation economy, from which to bring alternative cotton. Concerning Egypt, two related questions were particularly salient for the experts who deliberated on structural questions of cotton expansion: scarcity of labour and the mechanisation of cotton growing and ginning through introducing more capital.

Already before the war started with the battle of Fort Sumter on 12 April 1861, a francophone Egyptian expert was thinking about how to end the global cotton economy's dependence on America. On 1 March 1861, John Ninet, the Swiss farmer we met previously as a prominent agricultural advisor to Mehmed Ali since 1839, published a paper about

57 Ibid.
58 Ibid.
59 Ibid.

'The Cotton Question in England since the American Crisis' in *Revue des Deux Mondes*. 'Who would have imagined that the prosperity of such a colossal industry as a whole [. . .] could depend on a revolt of slaves?' Ninet asked.[60] How did England, he added, leave the raw material for the industry that formed 'the marrow of its bones' in the throes of perilous chances?[61] Looking for alternatives for growing cotton across the world, and concerning the lands of the Ottoman Empire, Ninet thought the situation in Syria, Asian Turkish lands, and Anatolia desperate in terms of labour availability. Egypt had more potential but with some of the same limitations of labour: cotton-suitable land was available, and was capable of producing treble the amount of the half-million cantars it produced throughout the 1850s, only if 'free workers, blacks, or coolies' were imported to work on extended cotton plantations.[62] Especially if other Ottoman lands were sought after along with Egyptian lands, the European powers would need to supervise and protect, through deploying large battalions, the migration of such workers from Europe to the Orient, otherwise it would not work out.[63]

Concerning Egypt in particular, the most promising land with cotton-expansion potential among the countries of the Mediterranean as Ninet asserted, maximum capacity with native assets had already been reached by the 1850s. Even if it was possible to expand in growing cotton, 'the number of children available for picking' was so small that 5 per cent of the cotton balls in the fields already growing was left unpicked. But the industry could expand in Egypt, and capitalists would jump on the opportunity to handle the expansion if 'an addition of foreign arms, Chinese or Malay, would gradually develop the endless resources there'.[64]

Nothing in Ninet's focus on child labour was out of the ordinary. In fact, cotton production in Egypt well into the twentieth century hinged on child labour, which became subject to nationalist discourses on developing native capitalism.[65] Coolies, the Indian and Chinese

60 John Ninet, 'La Question du Coton en Angleterre depuis la Crise Américaine', *Revue des Deux Mondes* 32:2, 1861, 196–222, 196.

61 Ibid.

62 Ibid.

63 Ibid., 214.

64 Ibid.

65 Aaron Jakes, 'Boom, Bugs, Bust: Egypt's Ecology of Interest, 1882–1914', *Antipode* 49:4, 2017, 1035–59.

labourers forcibly migrated for indentured work on colonial plantations elsewhere across the second half of the nineteenth century, were another story.[66] They were likely the subject of experimentation in connection with cotton in Algeria in the 1850s, colonised by France by then. An anonymous French letter sent to the *Times* correspondent, dated 31 December 1860, which supposedly expressed the views of the French government at the time regarding Chinese labour, started by mentioning a Franco-Chinese treaty. According to the letter cited, the treaty was conducted with the view of, 'no doubt, obtaining a supply of labour for the cotton lands in Algeria'.[67] French experiments on growing cotton there resulted in 1855 in a tiny amount of good cotton, 'but the want of an organised system of labour similar to the slave system in the States caused the culture to be abandoned for a time'.[68] Meanwhile, 'imported' adult Chinese workers suffered a high death rate. So 'a plan has been proposed to import boys and girls [. . .] under the care of priests and sisters of charity' in French missionaries in China, that is, child coolies. Those priests and sisters of charity 'will cleanse and clothe them, and begin immediately a religious and secular education'.[69] When they were to be brought to Algeria, the boys and girls were to 'be ready with their little fingers to pick the cotton balls as they ripen', with the help of steam ploughs and improved horse-driven hoes, so as to grow cotton on a huge area 'at half the cost of the American [cotton], owing to the difference in the value of land and slaves'.[70] The children, reportedly, were to become apprentices for twenty years, always under the supervision of their missionary friends and tutors. They were also to be taught how to

66 Coining the practice as 'extractive rawfaring' (using humans as disposable raw material), Mark Driscoll traces the origination of forcing Chinese and Indians into labour (coolies) to the juncture of the introduction of Anglo-American capitalism to Southeast Asia in the 1830s, preceded and followed by primitive accumulation through drug-trafficking and the First Opium War, and the abolition of slavery across the extractivist capitalist colonies of the globe. Driscoll argues convincingly for reframing the story of the rise of developmentalism in Southeast Asia as a criminal process led by 'white narcos' who flooded China with opium then harvested humans (the coolies) as cheap labour and a profitable trade in itself, across the colonies. Mark Driscoll, *The Whites Are the Enemies of Heaven: Climate Caucasianism and Asian Ecological Protection*, Durham, NC: Duke University Press, 2021.

67 Cited in 'Chinese Coolies and Cotton', *Dublin Evening Mail*, 9 January 1861.

68 Ibid.

69 Ibid.

70 Ibid.

grow their own food and weave their own clothes while in Algeria, then after twenty years they were to become citizens, with allotted subsistence plots, or have the option of returning to China.

The plan, according to the letter, was to show that there was no scarcity of lands for growing cotton across the world to end dependence on the slave-grown cotton of America, a fact that was destined to be proved by 'the application of the steam plough with the coolie emigration'.[71] It is likely that Ninet was inspired by such endeavours and plans when he brought the Malay and Chinese coolies to his essay.

The British experts, instead of direct involvement through bringing in and supervising coolies to fix the labour shortage in Egypt (that is, imitating the slave-plantation cotton economy of antebellum America on the Nile), were thinking of indirect solutions through British steam technology and capital. According to the *Cotton Supply Reporter*, the Manchester Cotton Supply Association's periodical, between 1858 and 1872, steam technology supplied through foreign capitalists was the recipe for a limitless increase in production in Egypt:

> A country which has evinced thus practically its capacity of producing invaluable textile, deserves *more attention at the hand of foreign capitalists* than it has hitherto received. Were its fine *natural resources stimulated by scientific appliances, which capital alone can furnish*, there need scarcely be any limit to the supply of excellent cotton obtainable from thence.[72]

In other words, combine Egyptian nature with fossil capital and you will resolve the problem of the blocked American cotton supplies.

With the onset of the American war, the secretary of the Manchester Cotton Supply Association, Mr G. Haywood, rushed to Egypt in August 1861, on his way to India. When he met Sa'ed, the latter had on his mind more systematic penetration of foreign capital as the anchor for a sustained increase in Egyptian cotton produce. Sa'ed Pasha pitched the idea of establishing a 'cotton bank' with English capital, and full backing

71 Ibid. It is noteworthy that steam ploughs were declared successfully developed and started to be deployed on British soil only by the late 1850s and early 1860s. See Chapter 6.

72 *Cotton Supply Reporter*, reproduced in 'Progress of Egyptian Cotton Cultivation', *Glasgow Herald*, 21 July 1862. Emphases added.

of his government, to advance money to cotton growers at a 1 per cent monthly interest, replacing the 3 to 4 per cent charged by the money-lenders.[73] His exchange with Haywood did not show to what degree he was keen on the steam-powered cotton gin in particular, but we can deduce more concretely that he thought that lower interest rates on loans to the peasants would incentivise expansion in growing cotton on more of their plots while making for more regular payment of land tax. Haywood's long letter of assessment to the Viceroy after his survey of the cotton districts addressed the cotton bank idea and documented advances in fossil ginning since Fowler's report of 1859–60.

On his tour, Haywood had the impression that, while 'immense tracts' of land were uncultivated, and others were indifferently cultivated, given the limited capital invested in them, and where, on both, more capital could have put the land to better use, overall, 'a very much larger extent of land is under cultivation [in cotton] this year [1861] than during any period since cotton has been cultivated in Egypt'.[74] It indeed paid off, for the cotton exported from Egypt in 1861 was 20 per cent above the half a million cantars average of the 1850s (at 600,000 cantars in 1861), then reached 820,000 cantars in 1862, then far surpassed the 1 million threshold starting in 1863.[75] Some of the cotton fields were more productive, in quantity and quality. With the replacement of oxen-powered *saqiyas* with steam irrigation pumps on his mind, Haywood talked about 'modern agricultural implements' that would 'both save much time and labour – or in other words, increase their [i.e. peasants'] limited capital', hence increasing overall cotton production.[76]

Before him, Fowler had advocated the systematic use of steam pumps and proposed a detailed scheme for introducing more of them on the estates of the Viceroy, and at least one at each district, to educate the peasants about their utility so they would buy more pumps on credit. In fact, this was one of Fowler's three recommendations for increasing cotton production in Egypt: 'the only efficient substitute for the Sakkiah and the Shadoof would be, of course, a steam pumping machine', which,

73 Owen, *Cotton*, 78–9.

74 'Cotton Culture in Egypt', *Public Ledger*, 3 October 1861, reproducing in full Haywood's letter to Sa'ed Pasha.

75 Owen, *Cotton*, 90, Table 7.

76 'Cotton Culture in Egypt', *Public Ledger*, 3 October 1861.

while consuming expensive coal, operated at only 'a little over the cost of manual and animal labour' in irrigation.[77]

Concerning the other steam implement that Fowler had advocated, the mechanical gins, Haywood saw that they had already that summer reached the capacity of 600 to 700 steam-powered McCarthy machines, more than double the 300 gins Fowler learned about a year or two earlier.[78] While the first 300 were mostly the property of European entrepreneurs, the new ones included many that were owned by the Pashas. Moreover, Haywood saw 'well-built [ginning] factories in various districts, and some in course of preparation for cleaning the cotton crop', with the countrywide capacity for steam-ginning 25 per cent of the crop, after it stood at 12.5 per cent in Fowler's report.[79] Given the then-expected cotton supplies shortage in England in the short term, Haywood also suggested, beside pumping in more steam-powered gins, the introduction of enhanced 'hand or treadle gins', a model of which soon was to be sent to the Viceroy from the Cotton Supply Association.

For all these improvements – mostly fossil-powered – more money was needed for the purchases. Haywood declared his full endorsement, in principle, of Sa'ed's cotton bank scheme that needed British capital. He had telegraphed the idea of the bank 'or some similar organisation' to his 'constituents', and was already 'informed that men of influence in Alexandria itself are already prepared to cooperate in any well-devised plan for carrying this important suggestion into practical effect'.[80]

Yet Haywood had at least two strings attached to this enhanced British credit scheme, which he subtly revealed. First, he presented to the Viceroy 'the fact which I have been most desirous to be informed upon, namely, the cultivation of the American or New Orleans class of cotton' in the Delta, which he found through the tour to be 'eminently adapted to the soil of Egypt'.[81] The demand for it on European markets was growing, compared to Egyptian strains; it was 'unlimited'. He wanted His Highness and 'the great landed proprietors' to test this seed, before the European markets were cut off from American supplies over the

77 Fowler, *Report*, 20–1.
78 'Cotton Culture in Egypt', *Public Ledger*, 3 October 1861.
79 Ibid.
80 Ibid.
81 Ibid.

next two or three years.[82] Second, in combination with this plan for enhancing the quality and shifting to American types (never realised), Haywood added that 'there is ample room for the introduction of large number of overseers and managers, *having practical knowledge of agriculture on modern principles*'; British experts were requested to arrive with the American seeds and the British capital.[83] Then he said it bluntly: the cotton bank desired by the Viceroy was to be fashioned along the lines of a mini–East India Company in the Nile Delta:

> The establishment of company [not bank] under the auspices of Your Highness, for the introduction of capital at a reasonable rate of interest, might go far to remedy many of the wants I have already pointed out. Such an association *by its agents*, who must be men of high integrity, would speedily discover all, and many more than those deficiencies already adverted to, and, if properly supported under an efficient managing directory, would in the course of a few years bring about a great revolution and improvement in the agriculture and cotton trade of Egypt.[84]

The secretary of the Manchester Association further promised Sa'ed to induce through this English-Egyptian cotton-bank-turned-company the transformation of Egypt from 'one vast farm' into 'one vast garden', making 'her people rich and her treasury inexhaustible'.[85]

Further cotton commentary in the British press shed light on where Haywood's idea came from: the wish to secure the more extensive British financial capital desired to go over to Egypt. For example, in the same news story that announced Haywood's arrival in Egypt and his planned meeting with Sa'ed, a commentator said that replacing the 'Greek and Levantine' traders' onerous small loans with British capital needed some serious enforcement and 'superintending' from the

82 Ibid.

83 Although the Manchester Association sent seeds to the Viceroy and many big cotton producers after Haywood's visit, with instructions on how to grow them, these American seeds never produced a shift away from *Jumel* and other Egyptian cottons. See Owen, *Cotton*, 98. 'Cotton Culture in Egypt', *Public Ledger*, 3 October 1861. Emphasis added.

84 Ibid. Emphasis added.

85 Ibid.

Egyptian government.[86] He added that such an idea was probably not entertained by the Cotton Supply Association, but Haywood had, in fact, a few days later, suggested his British-Egyptian superintending and enforcement we saw previously. Then, in the summer of 1863, a commentator replying to Samuel Smith's cotton survey of Egypt from the *Liverpool Post* voiced concern over the precarious capital investment conditions in Egypt when compared to India. 'We must remember that trade is there exposed to much greater risks than in India, and that we are not likely ever to bring English influence to bear upon the conduct of the Pashas, in the same manner, or to the same extent, as we reasonably hope to do in our Indian dependencies.'[87]

In other words, according to this opinion, which might have voiced the concerns of many capitalists who were comparing Egypt to India at the time, if British capital was to be deployed in Egypt on a large scale, conditions there needed to be ameliorated in a direction that was to move the country closer to the British power setup in the subcontinent. The discussion was, perhaps, implicitly about eventually turning Egypt into a British company-state.[88] In hindsight, we can alternatively treat this tentative proposal as a precursor for the phenomenon of the Colonial Cotton Campaign that mushroomed in tandem with the scramble for Africa over the next decades. In 1896, German cotton capitalists formed the Kolonial-Wirtschaftliches Komitee (Colonial Economic Committee) to promote cotton production in Germany's African possessions. Also in sub-Saharan Africa, the British Cotton Growers' Association was formed in 1902, then in 1903 came the French Association Cotonnière Coloniale. Similar entities were established by cotton capitalists from Italy, Belgium, and Portugal in order to pump more cotton exports from sub-Saharan territories.[89]

But the cotton bank, company-state, or early semi-Colonial Cotton Campaign on the Nile was never realised. Sa'ed died in January 1863, and

86 'Egyptian Cotton', *Belfast Mercury*, 14 August 1861.

87 'Egyptian Cotton', *Saint James's Chronicle*, 13 June 1863.

88 Further on 'company-state' as a concept descriptive of the history of the East India Company, see Philip J. Stern, *The Company-State: Corporate Sovereignty and the Early Modern Foundations of the British Empire in India*, New York: Oxford University Press, 2011.

89 See detailed analysis on the rise and fall of the Sub-Saharan African Colonial Cotton Campaigns in Ross, *Ecology*, 44–66.

his successor Ismail turned his attention to steam-powering agriculture on an unprecedented scale for motives and factors and through means that we will examine in the following chapters. What concerns us thus far is this subtle hinting at the form taken by the greater British capital involvement in Egyptian cotton cultivation and ginning – from the representative of Manchester cotton interests – at the outset of the cotton famine/boom. For it shows how deeply entangled European fossil and financial capital in the form of ginning mills, irrigation pumps, credit schemes, and loans was with the monetisation of the agrarian economy in Egypt and the liberation of trade under Sa'ed, with cotton pulling the wagon. In other words, the transition from manual and animal energy to steam energy on the eve of the cotton boom in Egypt was imagined and sought after, along a trajectory of direct British cotton-based financial colonisation of large tracts of land in the Delta, through the never-realised bank/company, its agents, and shareholders of peasant debts.

Conclusion: Steam Technology and Loans

The scheme suggested by Haywood was a road not taken for the transition to steam energy in Egyptian agriculture, but the transition happened eventually over the second half of the 1860s and the 1870s, leading through a longer detour to the same destination: the fall of Egypt to British military occupation starting in the summer of 1882. Let us stop at what we can gather about links between the rise of steam-powered technology on the Nile so far, and loans.

One can read the longer trajectory as it unfolds for us into the early 1860s as that of the gradual rise of fossil capital in Egypt since the 1840s with the free trade it forced through. Short of fossil capital, Mehmed Ali's frustrated empire that, if successful, was to provide a substantial measure of independence – and to build another grand chimney beside that of Britain in the early nineteenth century – gave way to growing surrender to dependency on the core.

Without factoring in fossil capital, the literature brings forward some nuances of this epochal change extending between Mehmed Ali's decades and that of Sa'ed (1854–63). Capitalism taking hold is seen as a metamorphosis of Egypt from the Ottoman resource management and economic system – monopolies combined with extra-economic coercion of the peasants under Mehmed Ali – to a commodity frontier that became attractive to the British-centred imperial world economy.[90] Alexandria turned in the 1850s into a bridgehead for soft forms of imperial capitalist control and power (expanded British-centred trade, migration, investment, banking, and consular power) over Egypt, prefiguring the 1882 British occupation.[91] Then there is the useful view allowing for spotting continuity in how pre-capitalist extra-economic relations of production formed the bone marrow of capitalism in Egypt under those formative years. Starting in the 1850s the hybrid of pre-capitalism/capitalism changed with the Levantine joint-stock companies and their competition, and partnerships, with European capital, in association with the newly formed Egyptian landed upper class.[92]

Now, let us factor in the steam-powered cotton gin, the machine at the forefront of the proliferation of steam-powered technology in mid-century Egypt. It was the disruptions to the raw cotton supplies from the antebellum American South, which had been powered by 3 to 4 million slaves on the plantations, that pushed Egypt in the early 1860s to go over the limit of half a million cantars of cotton annually into double and then quadruple that amount within a few years. While the manual-labour pool was stretched to the limit, we saw how the gradual increase in steam-powered ginning factories helped the cotton yield's great expansion.

First, by the turn of the decade, 12.5 per cent of the total product was processed by the steam-powered roller gins; then it was doubled with the outset of the American war to 25 per cent, even though the cotton crop more than doubled; and one year later (1863) it was up to one third of the crop. The fossil capital that moved over to establish and operate the ginning factories did more than just increase the cotton-ginning capacity. It greatly supported the expansion in cotton growing by compressing ginning-time to a fraction of the manual ginning-time, thus making it possible for the farmers who cultivated the inferior *Balli* cotton to extend the product to more land plots once they knew that the increased produce would not lie unginned in barns for long months.

90 Jakes and Shokr, 'Capitalism', 141.

91 Michael J. Reimer, *Colonial Bridgehead: Government and Society in Alexandria, 1807–1882*, Boulder: Westview Press, 1997, Chapter 1.

92 Kristen Alff, 'Levantine Joint-Stock Companies, Trans-Mediterranean Partnerships, and Nineteenth-Century Capitalist Development', *Comparative Studies in Society and History* 60:1, 2018, 150–77.

Simultaneously, the steam-powered cotton gins intervened in the micro-loans cycle that made cotton production possible for the growers, while they paid quarterly cash tax to the state. Mechanised cotton ginning decreased the time between harvesting and selling cotton for many farmers, helping them pay less monthly interest on the loans they took from traders and usurers. When British fossil capitalists noticed the great returns on investments in ginning on the Nile, they rushed to the Delta with their capital to establish those factories and made 20–5 per cent net profits.

Then British financial capital was itself sought after by Sa'ed Pasha as he saw British enthusiasm for the ginning factory. He wanted to replace the heavy-interest micro-loans that in principle enabled the state's monetary tax collection – and hindered it with the heavy rates imposed – with a more organised British-Egyptian cotton bank that offered more favourable loans to, at least, the smaller cotton growers: British capital and Egyptian state organisation and enforcement. The sought-after British capital came with intentions and tentative plans of greater control through loans and steam machinery. The Manchester Cotton Supply Association's secretary visited Sa'ed and suggested a company with overseers and supervisors instead of the bank. It was an enterprise imagined after the model of the British Indian colonial project, or an early plan of the European colonial cotton campaigns/companies that blossomed some decades later across sub-Saharan Africa.

It must be stressed, however, that capitalism does not seem here to have had any global master plans deliberately conjured for expansion. If anything, it operated in and perpetuated an anarchic state of affairs that was itself the fruit of that expansion. The steam-powered ginning factory owners from England and elsewhere in Europe came to the Nile to harvest the lucrative net profits of 20 to 25 per cent as they saw an opportunity. The Manchester Cotton Supply Association, as gleaned from Fowler's report of 1859–60, then Haywood's letter of assessment to the Pasha in 1861, was simply pursuing expansion in cotton imports to British factories – and further capital accumulation from the cotton before it was shipped to British docks – by complementing manual labour in short supply in Egypt with steam-powered ginning machinery. When the idea of extending credit to the Egyptian cotton agriculture came to the table, the suggestions put forward to realise the extension of credit to the Delta were based on how such schemes were handled thus far by the British capitalists in the Indian hinterlands.

The time-compression brought about by the ginning factories in Egypt in the early 1860s, which was eating away at those months previously required for the manual ginning of cotton, had other impacts as well. It released into the atmosphere carbon particles that, in the final analysis, subsidised the payment of cash taxes to the Pasha and subsidised the expansion in cotton production for the British mills already releasing many more carbon particles over the earlier decades. The ginning factories compressed the time needed for turning the cotton into a sellable commodity for money, to quickly repay the loans that the landholders took from usurers to grow the cotton before they accumulated more interest with time passing, and to pay the cash taxes Sa'ed imposed in place of in-kind taxation.

A Capitalocene dynamic thus was initiated, one in which it was imperative that steam power be deployed in the Delta, in the greater context of global cotton-based industrial and commercial circulations, and material (and financial) exchanges. It took hold and became the model on which further proliferation of steam power in later years continued. What was it, and how was it different from the logic of steam technology's proliferation before the cotton famine of the early 1860s?

The agrarian economy that Mehmed Ali sought to complement by fossil-capital-reliant industrialisation to serve his project of an independent predatory state (perhaps imagined by him to become a rival of Britain) floundered with the loss of monopolies and the sudden redistributions of the land (the measure he crafted to circumvent British forced free trade). The landed upper class – which, as it was being formed, contributed through the early 1840s cattle murrain to ending the Pasha's project – adopted steam technology in agrarian production for two reasons.

First, steam technology was deployed only as far as it manoeuvred the deteriorating perennial irrigation system (Cornish irrigation pumps). Second, as long as it sustained the production of the main cash crops where steam technology became in the 1850s an integral part of their standard global market quality benchmarks (long staple cotton that needed abundant steam irrigation and white sugar that needed refining by steam technology). What seems to be the most radical change thus far was the wholehearted monetisation of tax across the Delta starting in the mid-1850s, which Sa'ed decreed under the pressure of western European financial capital that was seeking markets outside its metropolitan centres, itself a process resulting from the mid-century European economic crisis. With it, European financial capital started to make its way across the cotton fields of the Delta, taking the form of fossil-fuelled technology for cotton ginning, which competed against and collaborated with circles of Levantine and local capital. Then came the global crisis of the cotton famine with the American Civil War, the effects of which were to intensify this trend and turn Egypt into a lucrative market for yet more European financial capital. Expected big returns on investment in cotton became, for a while, more solid, as long as the global market remained deprived of American cotton.

The Capitalocene's dynamic we are exploring here, then, seems to be one of expansion of fossil-fuelled technology through crises. The crisis that opened the doors for feeding Egypt-as-a commodity-frontier steam technology to produce the commodities in question was the cattle plague/redistribution of land in the 1840s. The crisis that brought steam technology to the Nile on an unprecedented scale was the cotton famine in the early 1860s, and, in its turn, it soon became entangled – in the spring of 1863, as we will see in Chapter 6 – with a second, even more devastating cattle plague on the Nile. Crises dashing in a long relay race, one handing the baton to the next, running forward towards a fully fossilised economy.

Here are some observations gleaned on links between the extension of fossil capital to new lands in association with loans, being processes that brought together the rise of the Capitalocene in Egypt. As this pair – steam-powered technology and loans – progressed dialectically, the carbon particles released kept increasing as fossil capital kept penetrating across Egypt, over the next two decades.

Over the later years of the 1860s and the 1870s, fossil capital took more forms besides the cotton gins and irrigation pumps: steam cultivators (ploughs), locomobiles (mobile steam pumps), then modern steam-powered sugar factories became the non-human protagonists of Ismail's reign of loans, bankruptcy, and the steam engine. We turn to him and his age next.

PART II
Combustible Carbon

The sucrose molecule, the chemical building block of white sugar, is a combination of one glucose molecule and one fructose molecule. The sucrose molecule is composed of twelve carbon atoms, twenty-two hydrogen atoms, and eleven oxygen atoms. $C_{12}H_{22}O_{11}$ is the formula. Coal, meanwhile, when analysed by weight, assuming it is dry and ash-free, has the same main components in different proportions: carbon (84.4 per cent), hydrogen (5.4 per cent), and oxygen (6.7 per cent), as well as trace amounts of other elements. Two very different and yet quite similar substances. One is white and consumed by living organisms as a primary source of energy. Inside living bodies, sucrose devolves into glucose, forming the only energy currency inside the human body: the ATP particles produced by the cells' mitochondria. The other substance is black or darkish brown and consumed by inanimate objects for the great amount of heat and kinetic energy its combustion releases. Coal is plant matter that metamorphoses by chemical processes over millions of years of burial under the ground, during which the biomass's chemical composition shifts towards higher energy efficiency, becoming a container of densely packed solar energy of limited quantity, which had been used in monumental amounts since the dawn of the Industrial Revolution.

These chemical similarities between sugar and coal were probably not a concern to Khedive Ismail when he imported factory upon factory for the production of sugar in the late 1860s and early 1870s. He acquired some of the world's most technologically advanced sugar production lines, but coal, until then, was prohibitively expensive for the sugar commodity production to be profitable on the Nile. He was, however, following the spirit of the age closely, and that included the generous use of the very plant for producing sugar, the sugar cane, as combustible agent in the steam engines that powered its processing into sugar. Sugar cane does not come close to coal in efficiency of energy storage, but it belongs to some of the most efficient plant species in terms of storing solar energy, with the ability to convert up to 1 per cent of the solar energy intake during photosynthesis into biomass. In the hard fibrous and dry matter (around 8 to 14 per cent of the plant is fibre; 11 to 18 per cent sugars; 65 to 75 per cent water) resides also great combustion potential.[1] This fibrous matter, also known as bagasse, is most of what is left after the crushing of the hard cane shoots to extract juice.

The Khedival sugar factories of the 1860s and 1870s were using bagasse to fuel their steam engines. Carbon of two varieties, edible and combustible, dominated the production cycle: the latter for production, the former as the product. The broadest line of this story of carbons lies outside the sugar factory. The mobile steam engines (locomobiles, or steam cultivators) found their way to the farm as prime movers in England in the late 1850s. By the early 1860s, these steam-powered machines – steam ploughs, and steam ploughs retrofitted as steam irrigation pumps – penetrated parts of the world where coal was expensive enough to make them a bad investment. The new production technology became available for the coal-deprived farmers of the world, but the unavailability of local supplies of the black carbon needed to power them stood in the way. As such, innovative solutions were invented, using other carbons besides coal, abundantly available where plants grow. Ismail's Egypt (1863–79) was a central node in this process of practical knowledge generation and harnessing for commodity production outside the global centre of coal, the British Isles.

In the chapters that follow, we cover what the steam engine, and its fuels, did to the Nile valley and its people, across the Delta and in Middle and Upper Egypt. As such, Part II historicises what was likely the most substantial emergence of steam-powered technology in agrarian production anywhere across Africa until the 1870s.

1 H. Savastano et al., 'Sustainability of Vegetable Fibres in Construction', in *Sustainability of Construction Materials*, edited by J. M. Khatib, Cambridge: Woodhead Publishing, 2009.

5

The Age of Debt, Steam Engines, and Bankruptcy (1863–76)

Why did Khedive Ismail keep sinking into loans, one after the other, across his years in power – almost one loan every two years – and why did European financial capital keep granting him new loans even while it seemed illogical? These questions serve two purposes. First, to approach his entire reign, as an age of mounting debt, against the tide of interpretations reached by national historians, which aligned with the narrative of the British colonial project after 1882, that Egyptian governance was inherently inefficient.[1] This mainstream narrative claims that Ismail was a wasteful despot addicted to loans, who was deceived by European banks bent on exploiting and ultimately colonising the country by feeding this addiction. Second, this is where we set the stage to analyse the progress of fossil capital in combination with financial capital (another name for loans), the pair that formed together, across time, the extension of the Capitalocene to Egypt. As the following chapters will show, Ismail's decades in power witnessed a great and sudden transition towards steam power.

1 For example, Abdulrahman Al-Raf'i, *'Asr Ismail [Ismail's Reign]*, Cairo: Al-Nahda, 1948; Mohamed Sabry [Al-Sourbonny], *L'Empire Égyptien sous Ismail et l'Ingérence Anglo-Française (1863–1879)*, Paris: Paul Geuthner, 1933.

Loans as Bridges

When the next step of the steam energy transition happened under Ismail (1863–79), it was heavily subsidised by European loans that landed Egypt by 1876 into a total debt of no less than £90 million, including accumulated interest to shareholders. In 1864–73, the Pasha borrowed no less than £65 million from European banks, and after many deductions, only around £44 million reached the state treasury, during a decade when total annual state revenues averaged around £7 million.[2] The money spent on steam technology and associated public works for production and trade in cash crops, like extending railways across the cotton centres in the Delta and light railway for sugar in Middle-Upper Egypt, the Alexandria docks, and Suez Canal digging, accounted for the expenditure of most of that money.

The first loan, of 1864, was justified by Ismail as needed for expenditures accrued due to a disastrous cattle plague in 1863.[3] Contrary to Al-Raf'i's estimates – which aligned with British ones that alleged financial inefficiency on the part of the Egyptian government, to justify occupation in 1882 – the plague was indeed a major turning point in which the country pivoted to heavy reliance on hundreds of imported steam engines to replace animal power in agriculture. Hence, the first loan (£5.7 million) must have been strongly related to the transition to steam technology. The second loan (£3 million in 1865) was said to have been conjured by Ismail in haste as negotiations on the first one took more time than he expected.[4]

The 1870 loan, of £7 million, was described by the assets mortgaged to it as 'the Daira Saniya loan' – a reference to Ismail's extensive estates across Middle Egypt. And, alternatively, by its claimed purpose, 'the sugar loan', as most of it was spent on importing and overhauling around twenty sugar factories on the Daira Saniya estates south of Cairo, with British and French steam-powered technology, and on the light railway network quickly built to serve the factories.[5] Ismail turned to refined

2 Roger Owen, *The Middle East in the World Economy*, London: I. B. Tauris, 2002, 124–7.

3 Al-Raf'i, *'Asr Ismail*, 35.

4 Ibid.

5 Ibid., 44; emphasis added, regarding the 'Daira Saniya loan'. The 'sugar loan' was so described generally in the British press: 'Mortgage Loan of His Highness the Khedive of Egypt', *Sun*, 26 April 1870; no title, *Daily News (London)*, 7 May 1870; 'London, Wednesday, April 27', *London Evening Standard*, 27 April 1870.

sugar as the commodity that was to replace cotton – after the American war ended in the mid-1860s – as his winner crop on the international market.

Besides direct spending on steam machinery, in Roger Owen's tentative estimates, between £10 million and £13 million was spent on building railways between 1863 and 1875, across the Delta and as far south as Assiout, mainly to serve production and trade in cotton and sugar.[6] Around £12 million was spent on building new irrigation canals – including the Ibrahimiya canal dug in the mid-1860s to help expand in grains, to replace lands under cotton in the Delta, then some years later for growing sugar cane – and maintaining and deepening existing ones, including the summer canals for perennial irrigation needed for both cotton and sugar cane.[7] The Egyptian government had also lost around £16 million to the Suez Canal digging project, including heavily subsidising the importation (from France) of a large cluster of steam dredgers, to replace manual corvée labour that started the digging in the 1850s, 20,000 labourers who worked on the canal at any given time until the mid-1860s.[8]

Overall, between half and three quarters of the European loans that led to state bankruptcy in 1876 were spent on steam technology and the infrastructures it needed – railways, canals, docks, and so on – the latter being necessary for realising revenues from such agricultural production and commodity transportation technology. This complicates things for the narratives on Ismail's addiction to loans. The loans spent on infrastructures to plant the technology across Egypt have rarely been seriously accounted for as a proper category worthy of historical analysis, even while such implantation was the purpose of the greater body of these loans.

The links between steam technology and European loans in the third quarter of the nineteenth century went deep. One could not progress without the other. We saw in Chapter 4 how Levantine and European financial and fossil capital intervened in the expansion of cotton across the Delta in the form of micro-loans and steam-powered cotton gins in

6 Owen, *The Middle East*, 128–9.

7 Money spent on railways, canals, bridges, telegraphs, Alexandria docks, and sugar: ibid., 131, Table 21.

8 Zachary Karabell, *Parting the Desert: The Creation of the Suez Canal*, New York: Random House, 2004, Chapter 16.

the early 1860s. But what lies outside the realm of a single commodity? How did capitalism in Egypt, and its fossil assets, develop under Ismail's years in general, if there is a general trend to discern?[9] In the passage that follows I present some mid-range theoretical interventions that speak to these questions, on transitions and dynamics of global capitalism, fossil capital, and ecological aspects of the movement of financial capital. They fit as interpretative tools in the broader theoretical scheme of the logic of gradation, the landscape of pre-capitalism/capitalism I presented in the Introduction. I begin with one of the most serious macro-economic analytical engagements with Ismail's loans that came very early in the twentieth century. Rosa Luxemburg presented Ismail's reign and his loans and economy as a case study for her hypothesis on the ultimate realisation of capitalists' surplus value through 'international loans' to pre-capitalist economies: the imperialist phase of capitalist accumulation.[10] The process goes through two arguments.

Luxemburg's Second Argument: International Loans and the Accumulation of Capital

The first argument is that successive cycles of expanded reproduction essentially need non-capitalist buyers of the produced goods, for the surplus value that will be redirected to expanded reproduction to be first realised. This is an argument that dealt extensively with Marx's theoretical formulae on expanded reproduction in *Capital, Volume II* as a way for Luxemburg to prove in abstract terms the necessity of outside markets for commodities produced by capitalists. The second argument (let us call it the empirical, historical, open-ended argument) says that it appears historically that cycles of expanded reproduction – turning realised surplus value into productive capital accumulation – tended to

9 'Capitalism in Egypt' as distinct from 'Egyptian capitalism', a distinction stressed in Aaron Jakes and Ahmed Shokr, 'Capitalism in Egypt, Not Egyptian Capitalism', in *A Critical Political Economy of the Middle East and North Africa*, edited by Joel Beinin, Bassam Haddad, and Sherene Seikaly, Stanford: Stanford University Press, 2021.

10 Rosa Luxemburg, 'The Accumulation of Capital: A Contribution to the Economic Theory of Imperialism', in *The Complete Works of Rosa Luxemburg, Vol. II: Economic Writings 2*, edited by Peter Hudis and Paul Le Blanc, translated by Nicolas Gray and George Shriver, London: Verso [electronic version], 2015, Chapter 30.

happen through exchanges with and expansion into pre-capitalist environs surrounding the capitalist economic formations.[11] While the first argument was criticised as underconsumptionist and essentialist, the second one passed the test of scrutiny. Put in one sentence, according to Tony Brewer: 'If Luxemburg was wrong to argue that capitalism needs a non-capitalist environment, she was surely right to stress that it did in fact appear and grow in such an environment.'[12] David Harvey too, rejecting the underconsumptionist logic running through her first argument, saw her second one as of interest, then tweaked it to 'capitalism necessarily and always creates its own "other"'.[13]

To salvage the open-ended historical second argument, Brewer suggested that we can read Luxemburg with an eye for replacing essentialism with potentiality.[14] We can, indeed, understand her analysis of the realisation of surplus value in pre-capitalist societies, and the transformation of realised surplus value into expanded productive capital, as a possibility not a necessity. This suggestion can be usefully applied to the following generalisation of Luxemburg: 'Be that as it may, the fact of the matter is that the accumulation of capital as a historical process, in all its relations, *is* [read: tends to be] contingent upon noncapitalist social strata and forms.'[15] What concern us the most here, after this disclaimer on necessity, are the processes through which, according to Luxemburg, European capital penetrated pre-capitalist formations, simultaneously being characteristics of the imperialist phase of capitalist accumulation: international loans, building railways, revolutions, and wars.

How these processes might work together, in short – and let us take the example of England and Egypt – is as follows. Exchanges of international trade in raw commodities (cotton) and finished commodities (textile exported back to Egypt) ultimately realise surplus value in pure value-form (money) that goes to banks and commercial houses in

11 Tony Brewer, *Marxist Theories of Imperialism: A Critical Survey*, London and New York: Routledge, 1990, 58–9, paraphrasing Luxemburg's theory on accumulation (Chapter 26); also in David Harvey, *The New Imperialism*, Oxford: Oxford University Press, 2005, 137–41.

12 Brewer, *Marxist Theories*, 66.

13 Harvey, *The New Imperialism*, 140–1.

14 Brewer, *Marxist Theories*, 64, following Joan Robenson's arguments in her introduction for the Agnes Schwarzschild translation of *The Accumulation of Capital*.

15 Luxemburg, 'The Accumulation', Chapter 26. Emphasis added.

England. Then the pure value-form fulfils its role in expanded reproduction by going back to Egypt in the form of international loans, which are then used to establish productive capital – including steam-powered production technology – railways, and other infrastructures for the new productive capital on the Nile. Then compulsion in the form of war unleashed on Egypt in 1882 by Britain served the purpose of facilitating further accumulation, now that the Egyptian pre-capitalist economy became tied closely to British capital, and after it declared bankruptcy under the impact of excessive loans.[16]

Zooming in on the question of loans, Luxemburg suggested that it supplied money for buying 'machinery, materials and the like', items simultaneously produced by and exported from the country from which financial capital comes. With this, 'the new [pre-capitalist] countries comprise huge regions characterized by natural economic relations, which are transformed into relations based on the commodity economy, or by relations based on the commodity economy, which are displaced by capital'. The loans also effectively exploited native labour ('the indigenous workforce of the new countries [. . .] is forced into submission by the capital of the old countries') and becomes 'the new object of its exploitation'. Meanwhile, labour from the loan-issuing old country joins forces with the native workforce, especially in the first stages where technical expertise is wanting for running the productive capital bought by and exported from the old capitalist economy.[17]

The conventional historical narrative of Ismail's wastefulness and European financial capital's conspiracies and ploys against Egypt over the 1860s and 1870s in the form of irrational successive loans finds a rational fine-grained alternative through Luxemburg's second argument:

> from the standpoint of capital accumulation, there was a perfectly rational and 'healthy' relation underlying the ostensibly absurd transaction between European financial capital and European industrial capital, in which the former provided the finance for the Egyptian orders placed with the latter, with the interest payments on one round of borrowing being serviced by the next. Stripped of all obscuring

16 Ibid., Chapters 30–2.

17 All citations in paragraph: ibid., Chapter 30.

> mediations, the underlying relation consists in the simple fact that the Egyptian peasant economy was consumed by European capital on a huge scale: enormous stretches of land, vast amounts of labour-power and masses of the products of labour, which accrued to the state in the form of taxes, were ultimately transformed into European capital and accumulated.[18]

Overall, Luxemburg's second argument, combined with her empirical analysis of Egypt's case, is suggestive of how international loans for buying steam engines and their infrastructures can help reinterpret Ismail's reign, as a crucial moment in the transition to steam power in nineteenth-century Egypt, contrary to the conventional narrative that propagated the hypothesis of the extravagant oriental despot.

I do not claim that this history was inevitable. On the contrary, the possibilities available for European capital investments, actively looking for buyers and borrowers, were simply realised in Egypt. For example, the 1870 loan for sugar was sold to British bondholders – in bonds of £20 and its multiples of £100, £500, and £1,000 – and they bought the loan shares purely as an attractive financial capital investment opportunity at 7 per cent annual interest paid biannually.[19] In an anarchic capitalist social formation, the loan was collected from all sorts of buyers, big and small. In other words, capitalism has no collective mind pushing it to expand into pre-capitalist formations, or pre-capitalist/capitalist landscapes. Rather, a plurality of capitalist actors see the potential of realisation of surplus value and expanded reproduction in some pre-capitalist territories, so they jump on it. The argument works the other way: pre-capitalist actors see potential for further realisation of surplus value by making use of loans that buy technology and infrastructure (railways, bridges, and so on) for increasing the volume of international trade in their agricultural commodities, so they jump on it. The discussion on Luxemburg's *Accumulation* just covered speaks only of conclusions within this range of possibility.

18 Ibid.
19 'Mortgage Loan of His Highness the Khedive of Egypt', *Sun*, 26 April 1870.

Globalised Fossil Capital and Cash-Crops Production

In historical analysis that accounts for its emergence, what theoretical tools do we have to take Luxemburg's piercing analysis on the expansion of capitalism to non-capitalist social strata and forms of social organisation closer to the realm of fossil capital?

As previously stated, I look at the arrival of the Capitalocene in Egypt through the progress of the movement of the pair of steam technology and loans, which brought colonialism or, more precisely, growing dependency followed by military occupation. Whereas the discussion on Luxemburg brought to sharp relief financial capital and colonialism, the few insights to come factor in fossil capital and loans more substantially.

Addressing the moment of accelerated globalisation since the 1970s, Andreas Malm proposed a theory of globalised fossil capital, according to which '*globally mobile capital will relocate factories to situations where labour power is cheap and disciplined – where the rate of surplus-value promises to be largest – by means of new rounds of massive consumption of fossil energy*'.[20] In this hypothesis speaking to an entirely different moment from ours, Malm hangs great importance on the availability of infrastructures, of energy sources to power the factories and transportation networks to move production assets and commodities. He proposes that globally mobile capital will always look to the state to provide such services before a country is attractive enough for coupling this global capital to carbon emissions through factories.

Now, what Ismail Pasha did was transform the agrarian economy more intensively than his predecessors towards cash-cropping, building a modern fossil-powered infrastructure for its production and transportation. He built an extensive railway system across the Delta and Middle Egypt, using £10 million to £13 million (25 per cent of the combined European loans). He spent around £12 million more (another 25 per cent) on building new canals. The Ibrahimiya canal in Middle Egypt, just one of these new canals, was said to be the largest and most extensive project of its kind anywhere in the world across the nineteenth century. Bringing steam technology to cash crops involved steam ploughs and irrigation pumps in the early 1860s mainly to grow cotton in the Delta

20 Malm, *Fossil*, 333. Emphasis in original.

and importing around twenty fully steamed sugar factories to his expansive Daira Saniya (a Viceroyal latifundium of sorts) across Middle and Upper Egypt in 1870. The sugar alone consumed no less than £6 million, or roughly 13 per cent of the combined European loans he took until 1876. He was indeed building an unprecedented infrastructure, but it was not with the intention of bringing in foreign investments to build and operate factories in the contemporary postcolonial sense. It was mainly to serve his own extensive personal land possessions, for, during these years, his own lands expanded to 20 per cent of all cultivated land in Egypt. He was indeed building a great plantation. During these years, the overwhelming majority of European capital moving to Egypt took the form of financial capital, beside the budding cotton-ginning industry that globally mobile capital had been investing in and erecting its factories for even before Ismail, from the 1850s onwards. Ismail's spending, through the loans, was to finance the refurbishing of the agrarian infrastructure and the introduction of steam-powered technology for growing cash crops. While the steam technology bought was not necessarily deployed by European capitalists, it was itself a commodity manufactured by some of them. Without reducing this process to a partial claim, that some European fossil capitalists – engine and machinery manufacturers – sold their commodity to the Egyptians using the loans, and stopping the analysis there, how can we square it with Malm's commonsensical hypothesis on the globalisation of fossil capital?

We can take our cue from deliberations on how capitalism came historically to dominate agriculture, which is also to elaborate on Luxemburg's claims about how capitalism swallowed up the Egyptian peasant economy. In 1925, A. V. Chayanov hinted at a categorisation of two ways through which peasant farms are drawn into 'the system of the capitalist commodity market': first is the latifundia category, the big capitalist farming model, and second is the capitalist 'economic dictatorship over considerable sectors of agriculture'.[21]

Henry Bernstein took this insight forward, and others gleaned from Lenin in *The Development of Capitalism in Russia*, to formulate a dual taxonomy for the metamorphoses of historically witnessed pre-capitalist to

21 A. V. Chayanov, *The Theory of Peasant Economy*, edited by D. Thorner, B. Kerblay, and R. E. F. Smith, Homewood, IL: The American Economic Association, 1966, 49.

capitalist social formations in agrarian settings.[22] This penetration of capitalism into agrarian economies is either the 'classical model', also known as 'the horizontal concentration of means of production (land, machinery, and labour-power) in units of production equivalent to industrial enterprises in their organisation of production and modes of economic calculation.' Or it takes the form of vertical concentration, which 'refers to the coordination, standardisation, and (greater or lesser) supervision of the production of numerous individual small producers' by an overarching entity.[23]

Based on this categorisation, it is reasonable then to see in Ismail's expanded agrarian economy, boosted by infrastructure for transportation and steam technology, the formation of the horizontal concentration of the means of production type, the latifundia or classical model of capitalist organisation of agriculture. In order to make it happen, however, Ismail relied heavily on global financial capital. Ismail's latifundia economy – the horizontal concentration of means of production – collapsed by the late 1870s, when the Khedive was struggling to meet his loan obligations, then state bankruptcy was declared in 1876, and he lost power in 1879 to his son Tawfiq. The most prominent example of this horizontal concentration of the means of production is the sugar factory, which rose and fell in the blink of an eye in the early 1870s. The infrastructure built for it, however, remained functional. It was in the early 1880s, with the British military occupation, when a certain vertical concentration of the means of production started to take hold. This occurred through direct European capital involvement in agrarian Egypt going deliberately through colonial governance to small farmers through small loans, and through colonial agrarian policies that sought to boost the productivity of the fellah (the small farmer), by building on the infrastructural networks Ismail started two decades earlier.[24]

22 Henry Bernstein, 'Notes on Capital and Peasantry', *Review of African Political Economy* 10, 1977, 60–73. The paper is organised in notes. Note 45 brings up the genealogy from Lenin and Chayanov. On this lineage and Bernstein's concept of vertical integration, see Jairus Banaji, 'Merchant Capitalism, Peasant Households and Industrial Accumulation: Integration of a Model', *Journal of Agrarian Change* 16:3, 2016, 410–31.

23 Bernstein, 'Notes', note 44.

24 Aaron Jakes revisited the mainstream narrative on the marriage between British colonial and big native business right after the occupation and presented historical material supportive of how the British colonial administration was, in fact, bent in its agrarian policies in the 1880s on making the small farmers (the fellah) more productive. See Aaron Jakes, *Egypt's Occupation: Colonial Economism and the Crises of Capitalism*, Stanford: Stanford University Press, 2020.

In other words, Ismail Pasha used loans – the deployment of European financial capital – to enhance and build the infrastructure necessary for cash-cropping along what Bernstein, after Chayanov, describes as the horizontal concentration of the means of production. At that moment in the 1860s, steam-powered technology in agrarian production was the protagonist around which this horizontal concentration was assimilated. It was the borrowed financial capital invested in this phase that struck back, like a boomerang, in the late 1870s, taking the form of occupation and colonisation, and more direct and indirect international investments using fossil and financial capital well into the twentieth century.

Combine these two consecutive phases – horizontal and vertical concentration of the means of production – with Malm's conceptualisation of the process of globalising fossil capital, and we will have an interpretation that elaborates on and takes forward Luxemburg's second argument on how the reproduction of capital from Europe in the form of loans swallowed up the Egyptian agrarian economy. Capitalism has a tendency to grow in pre-capitalist environments. It realised this tendency in the case of 1860s–70s Egypt through spending European financial capital in the form of loans in order to purchase and deploy steam-powered technology, primarily to produce the crop: cotton, then sugar when the cotton boom ended unexpectedly in the mid-1860s.

This is how I approach the historical analysis that follows. This approach demonstrates how the Capitalocene was extended to Egypt across the 1860s and 1870s through the pairing of steam-powered technology to loans that brought colonialism. It also serves the purpose of complicating the reductive mainstream narrative on Ismail's addiction to European money, that of the wasteful Khedive versus the inherently evil Europeans carrying out a premeditated conspiracy. The chapters that follow chart detailed accounts of particular moments of the transition to fossil-fuelled technology under Ismail's reign.

6

'The year of the terrible murrain among the cattle', the Year of the Steam Cultivator (1863)

The Rinderpest

Rinderpest, also known as the cattle plague, or the steppe murrain – which we encountered with Mehmed Ali Pasha – is a viral disease that for centuries killed cattle and wildlife across continents. In 2011, the World Health Organization declared it the second comprehensively eradicated viral disease in the world, after smallpox. The history of the early modern world and the Industrial Revolution carries the virus's heavy footprint. It killed 200 million animals in Europe across the eighteenth and nineteenth centuries. Most of the worst outbreaks happened across the eighteenth century, then the virus returned with a vengeance across Europe in the 1860s–70s.[1] A terrible outbreak occurred in England in June 1865, and it became a pivotal point of interest for the scholars of the history of British epidemics.[2] But the British 1865 cattle plague wreaked havoc elsewhere before it reached the Isles. In 1861 or

1 Johanna Widenberg, 'Cattle Plague and Society: Rinderpest, Anthrax and Epizootic Control in the Eighteenth Century Kingdom of Sweden', *Sjuttonhundratal* 17, 2020, 8–33. A recent and substantial world-ecological treatment of the 1860s cattle murrain: Chris Otter, *Diet for a Large Planet: Industrial Britain, Food Systems, and World Ecology*, Chicago: Chicago University Press, 2020, Chapter 1.

2 Terrie M. Romano, 'The Cattle Plague of 1865 and the Reception of "The Germ Theory" in Mid-Victorian Britain', *Journal of the History of Medicine and Allied Sciences* 52:1, 1997, 51–80.

earlier, it started spreading from Russia to the cattle of central Europe, before it reached Turkey in December 1862.[3]

In 1863, the plague found its way to Egypt. The outbreak started in the Delta around March 1863, and it became worse as Ismail imported more cattle. By the end of the year, the situation was thus described by Rev. Andrew Watson, the head of the American Presbyterian Mission in Egypt:

> This was the year [1863] of the terrible murrain among the cattle in Egypt. It is said that about 300,000 cattle died. Many of these were thrown into the Nile and the canals, polluting the water, and causing widespread sickness in the land, and preparing the conditions necessary for the terrible visitation of cholera in 1865. At one time the whole coast of the Mediterranean, in the region of Damietta and Rosetta, where the Nile flows into the sea, was piled up with carcasses of cattle in all stages of putrefaction.[4]

The archives of the British press between May and December 1863, along with a few Egyptian sources, show beyond doubt that the abrupt rise of steam power in agrarian production in Egypt was coterminous with the disastrous death of its cattle that worked the fields. We know that, on 22 January 1863, on his fourth day as Viceroy, Ismail Pasha ordered 1,000 oxen from Sudan 'to use in the works of cultivation and ploughing'.[5] Before the cattle came from Sudan, in March, the rinderpest devastated local populations in Behira, north-western Delta, and Ismail sent the army to impose quarantine by late May.[6] That month he also ordered oxen from Turkey, which was already heavily infected.[7] All cattle markets across the country were cancelled, but the outbreak spread across the Delta, killing three quarters of all oxen in a few days in

3 'Contagious Typhus in Cattle', *Farmer's Gazette*, 21 March 1863.

4 Andrew Watson, *The American Mission in Egypt: 1854–1896*, Pittsburgh: United Presbyterian Board of Publication, 1904, 160.

5 Amin Sami, *Taqwim Al-Nil 3.2 [Survey of the Nile Vol. 3.2]*, Cairo: Matba'at Dar Al Kutub, 1936, 447–8.

6 Plague in Behira: *Irada* (Viceroyal order) to Ismail Pasha the *Nazir Al-Jihadiyai* (minister of defence), in ibid., 481. *Times* piece: republished in several newspapers, including: 'The Plague Impending in Egypt', *South and North Lincolnshire Advertiser*, 4 July 1863.

7 'Egypt', *Cork Constitution*, 9 June 1863.

the late spring. *The Times* reporter who followed the progress of the plague suggested that Ismail should follow the lead of his uncle Halim Pasha and import 'steam ploughs and movable pumps to replace animal labour'.[8]

By then, a few months into his reign, Ismail had confiscated and bought huge parcels of land across the country, for which he ordered cattle. He annexed 200,000 acres in 1863, and by July 1864 he owned over half a million acres, or one ninth to one eighth of all cultivated land in Egypt, and the percentage was doubled to a fifth of all cultivated land by the 1870s, roughly 1 million acres. Some lands were bought from members of the royal family, but a substantial share was seized from peasants by coercion.[9]

The calamity did not stop at the murrain. Like the 1842 great cattle plague, a disastrously high Nile flood came on its heels, as Ismail scrambled for ploughing power of both kinds: oxen and steam engines.

In mid-August, the Nile inundation rose 4.25 metres above the one of 1862. It might have been the highest flood in a century.[10] The cotton crop was in danger, facing the double challenge of the rising Nile and the dying cattle. The outbreak had already reached Cairo. A certain Mustapha Pasha, likely Ismail Pasha's younger brother, lost 2,000 heads, worth £60,000, and some of the richest landowners suffered comparable losses.[11] The cotton-ginning industry was threatened with the loss of 25 per cent of its expected harvest.[12] The cotton shrubs reportedly needed to be uprooted on many estates and smaller farms and new seeds cultivated. Across many districts, no animals were left to run the irrigation wheels. Some fields were left without irrigation for two months. It became common, under these conditions, to see, 'for want of animals,

8 'Egypt', *Oxford Times*, 4 July 1863.

9 Roger Owen, *Cotton and the Egyptian Economy, 1820–1914*, Oxford: The Clarendon Press, 1969, 117, citing, among other sources, British Foreign Office correspondences FO 141/54, FO 78/1818.

10 Ibid., 100.

11 'Cotton in Egypt', *Liverpool Daily Post*, 31 August 1863.

12 Cotton production for 1863 was over 1 million cantars, for the first time in history. The production of 1864 and 1865 was even substantially higher, reaching over 2.5 million cantars in the latter year, meaning that if not for the murrain and the flood, production in 1863 might have become much more than 1.28 million cantars. See Chapter 4.

the fellahs [. . .] harnessed by fives and sixes to the plough'.[13] An eyewitness account from Luxor (Upper Egypt) talked of men dying in the spring of 1864 – when the plague went south eventually – of overwork as they replaced the oxen on the ploughs and the irrigation wheels.[14] In August, the *Sun* hinted at potential popular dissatisfaction with Ismail's newly acquired estates across the country. Popular anger was reportedly directed at both 'the question of the lands assumed by his Highness' and 'the proportion of forced labour' the peasants had to provide.[15]

Come September 1863, the Nile flooded Cairo, and Giza to the south. Four villages just north of the capital were completely swept away. While the flood receded quickly from the capital, in November the Delta was still suffering. That month, a British person was on board a steamer going from Alexandria to Cairo. At Tanta, Gharbiya, at the heart of Egypt's cotton fields, she saw:

> the spectacle of devastation – whole villages gone, submerged and melted, mud to mud, and the people with their animals encamped on spits of sand or on the dykes in long rows of ragged makeshift tents, while we sailed over where they had lived. Cotton rotting in all directions and the dry tops crackling under the bows of the boat. When we stopped to buy milk, the poor woman exclaimed: 'Milk! from where? Do you want it out of my breasts?' However, she took our saucepan and went to get some from another family.[16]

Eventually, the plague receded from Lower Egypt, after it found its way to Middle and, in a slower pace, Upper Egypt. That far, by September 1863, at least 200,000 animals had been lost, including as far south as Bani-Suef.[17] However, the cattle plague, combined with the devastating flood, increased 'sickness' among humans in Cairo and the Delta. This was likely because people buried many of the dead cattle close to their habitats. It cost 100 piastres to carry out governmental quarantine

13 'Egypt', *Cork Constitution*, 1 September 1863.

14 Lucie Duff-Gordon, *Lady Duff Gordon's Letters from Egypt*, New York and London: McClure, Philips & Co., R. Brimley Johnson (revised edition), 1902, 169.

15 *Sun*, 6 August 1863.

16 Duff-Gordon, *Lady*, 75.

17 'Egypt', *Bedfordshire Mercury*, 10 October 1863, citing *Le Constitutionnel*'s correspondent.

regulations and transport the dead animal for burial in the open, under conditions of 'famine' food prices, and an unusual level of labour scarcity. So, against the risk of sickness, many buried their dead cattle 'on their premises inside the town'.[18]

By early November, 300,000 heads had been lost, and at an average cost of £30 per head, they were to be replaced for £9 million, £2 million more than the average annual state income in mid-1860s Egypt.[19] It then rose to half a million cattle heads by mid-November, a loss of no less than £15 million.[20] By the last week of November, all kinds of mammals were reportedly dying. Although the Delta was relieved, Middle Egypt up to the southern borders of Minya suffered the most: cattle, donkeys, and camels were dying, the latter two perhaps not as extensively as the cattle.

The newly imported cattle that reached Egypt in the fall perished. Over the summer, Ismail sent his agents to Syria, Libya, Crete, Marseilles, Trieste, and Odessa – some of which were actually epizootic hotspots – to bring in new farm animals. After lifting customs on imported livestock, oxen, horses, and donkeys came through Alexandria. He sent some of these animals to his estates, and others were resold to rich landholders. It was suggested that the Viceroy resold most of the imported animals at prices slightly lower than those of livestock middlemen, which still compensated most of his losses in livestock that year.[21] Some of the newly imported animals, given the crisis, no doubt, were weak and small; others failed to adapt to the Egyptian climate, while some died from overwork. A second cycle of the plague was unleashed before the end of the year, and so Ismail sent orders again for more animals from his agents abroad.[22] By the first day of 1864, the country's cattle stock between the northern tip on the Mediterranean and Minya in the south was annihilated. By the spring of 1864, that of Upper Egypt too was totally consumed.[23] How did the common woman and man perceive this moment? In Cairo, around November 1863, in the midst of

18 Duff-Gordon, *Lady*, 78.

19 'Cattle Disease in Egypt', *Dunfermline Press*, 4 November 1863.

20 'Egypt', *The Globe*, 28 November 1863.

21 David Landes, *Bankers and Pashas: International Finance and Economic Imperialism in Egypt*, London: Heinemann, 1958, 191, note 2.

22 Owen, *Cotton*, 99.

23 Duff-Gordon, *Lady*. Progress of the murrain in Middle and Upper Egypt is discussed in Chapter 8.

the flood and plague, a fable most probably speaking of Ismail and his relentless importing of cattle kept spreading:

> The high Nile was a small misfortune compared to the murrain. There is a legend about it, of course. A certain Sheykh Al-Beled of some place lost his cattle, and being rich defied God, said he did not care, and bought as many more; they died too, and he continued impenitent and defiant, and bought on till he was ruined, and now he is sinking into the earth bodily, though his friends dig and dig without ceasing night and day.[24]

The conditions that made the epizootic what it was, a global disaster, were tied foremost to the context of global trade in cotton. If the world was moving in 1863 and the two years prior to frantically increase cotton production to make up for the slave-grown cotton of the US, then it is very likely that cattle raising – the animal energy necessary for growing that new cotton in lands with highly artificial irrigation setups like that of Egypt – was progressing relentlessly in tandem. More animals congregating in dense spots for raising and selling en masse to the new and expanding cotton producers of the world meant the greater likelihood of spreading the murrain globally, along the cattle trade chains. One can conclude, then, that the cotton boom caused the mid-1860s cattle plague more than any other cause.

In Egypt, the redistribution of land to the benefit of the new Viceroy, who appropriated for himself hundreds of thousands of acres over his first few months in power, also meant that he was keen to buy relatively large populations of cattle with less attention to their quality or health conditions. Officials of the Pasha abroad made rushed decisions to fulfil the urgent requests for animal power, and infected animals in this or that shipment's population showed up and the virus moved fast. In other words, what we have here – with the early 1840s murrain compared to the 1863 situation – is the tendency of rapidly expanded monoculture of energy-intensive crops to attract livestock epizootics. But unlike the epizootic of the early 1840s, when the virus – which became entangled, as in the 1860s, in sudden and extensive land redistributions – ended industrialisation by killing the animal power of the factories and shifted the rich

24 Ibid., 81.

landholder's thinking towards using steam engines in modest quantities, in 1863 the virus stimulated a sudden and frantic turn to steam power.

Agrarian production ached for prime movers. If the animals were sinking into the murrain vortex in droves right after they touched ground in Alexandria, then some other power had to be arranged. But, even before waiting to see the disastrous livestock importation results by early 1864, steam power, of virus-immune steel and coal, was arriving in tandem with the soon-to-die imported animals. Next we turn to the timeline of the 'steam cultivator' as it intersected with the timeline of the 1863 cattle plague.

The Invention of the Steam Cultivator

With the Industrial Revolution came steam cultivation. Between the turn of the century and the early 1830s, several pioneers laboured in experimentation to bring the Industrial Revolution to the farm. While steam engines that rarely moved were practical and economic to use for threshing around barnyards, the much higher mobility of the steam engine for ploughing across wet and irregular field terrain was a considerable challenge.[25] Over the 1830s and 1840s, three types of steam implements were finally settled on, identified as being the most promising.[26] Finally, in July 1858, the *Economist* reported that the moment for the steam cultivator had nearly arrived:

> There is a want of more power, and power in new shapes, to accomplish all in the way of cultivation our best agriculturalists desire to carry out. Hence plans for steam cultivation are eagerly regarded by so many of our 'fast' farmers.[27]

Tillage of the land in the autumn, with wetter soil and the need to plough deeper to prepare the land for root crops, required 'more horse

25 Clarence C. Danhof, 'Review: God Speed the Plow, the Coming of Steam Cultivation to Great Britain', *The Journal of Economic History* 21:2, 1961, 271–3; Reynold M. Wik, 'Review: God Speed the Plow, the Coming of Steam Cultivation to Great Britain', *Technology and Culture* 2:2, 1961, 187–9.

26 Danhof, 'Review: God Speed'.

27 'Steam Cultivation', *Economist*, 24 July 1858.

power than is commonly available – and horses are very costly machines'.[28] Consider for example the land dedicated to their fodder, which could have been freed up in their absence. There were high hopes therefore that steam cultivation could thus be a boon to British landowners.[29] It looked as though a steam plough that the average big landowner would find economical and practical – easy to move across the land, at costs lower than horse- and manpower – was in sight.

But it was coming in fits and starts: in 1857, the Salisbury meeting of the Royal Agricultural Society 'resulted in a report decidedly unfavourable to steam cultivators', but the inventors objected, saying that the testing ground was steep and difficult: it could be successfully tested but only if the ground was flat, with no fences or ditches.[30] Hope seemed to hinge on John Fowler's steam plough, which tilled the lands through the implements attached to it, which were moved by traction belts while the engine itself remained static.[31] Note this was not the same Fowler hired by Khedive Ismail the next decade for several irrigation and railways projects, nor the Fowler of Manchester we encountered in the cotton-gins economy of 1861 Egypt. This was Fowler of Leeds, an agricultural engineer who died in 1864. In 1861, a co-worker of Fowler described Leeds as the centre of English agricultural engineering, being in the middle of the coal and iron districts of Yorkshire. Mr Fowler was a 'young, amiable gentleman, a Quaker, the owner of emerging engineering works, the inventor of steam-ploughs [. . .] and a man who consorts with the whole world'.[32] In the late 1850s in England, his steam-powered traction belt machines were difficult to manoeuvre, due to the fact that the fences between English farms had all been designed around horses and humans. The fences needed to be taken down and the ditches needed to be filled if the heavy engine on wheels was to be moved from one ploughing line to the next. It seems that such innovation was more suitable for larger, flat, and fenceless plantations.

In August 1858, the *Economist* struck a more confident tone: 'The question of the practicability of steam cultivation has been solved.' The

28 Ibid.

29 Danhof, 'Review: God Speed'.

30 'Steam Cultivation', *Economist*, 24 July 1858.

31 Ibid.

32 Max Eyth, *Im Strom unsrer Zeit. Erster Band: Lehrjahre* [In the Flow of Our Times, Vol. I: Apprenticeship Years], Projekt Gutenberg-DE, 1904, Chapter 19.

Royal Agricultural Society awarded Fowler's plough engine the premium of £500 and recommended that a gold medal be given to another invention, one by Messrs Howard. With these good tidings from the Society, prospective investors in progressive agriculture were informed that it was certain 'that steam cultivators will greatly extend the power to till land in the autumn [. . .], and deeper culture will speedily follow'.[33] Many machines subsequently entered the Society's competition, and the judges applied several tests: ones related to daily working costs, and others for various kinds of soil. The results for Fowler's system showed slight cost savings over horsepower: 7s 2d (7 shillings and 2 pence) per acre for light land (compared to a minimum of 8s if using horses), and 9s 2d for heavy land (compared to 12s 6d for horses). Trenching in the tests cost 18s 4d per acre, and it was a task that horses could not do at all, the judges claimed. Instead, manual labour was needed there at over £6 per acre.[34]

While savings of up to 25 per cent for ordinary fieldwork through applying steam cultivation were not substantial enough to woo landowners into investing in the engines and their machines, the judges declared that 'when these machines are in constant work on a farm the cost will be less than that stated in the report, – while the engines are applicable to other power purposes of the farm'.[35] When the Yorkshire Agricultural Society held tests later, they awarded their prize to Fowler, not the Howards. In fact, it was said that Fowler won every competition he entered for showcasing and testing steam cultivators across the 1850s and 1860s.[36] The Royal Agricultural Society, by 1863, gave the agricultural press of England the impression that it had 'uniformly patronised Fowler's system'.[37] By then, nearly all of the Society's prizes related to the application of steam power to agricultural soil were granted to Fowler. He became 'steam cultivation' personified, while Messrs Howard and very few others came next.[38] Across the early 1860s, Fowler's name keeps popping up in the British newspaper archives in association with both steam cultivation and Egypt.

---

33 'Agriculture: Steam Cultivation', *Economist*, 14 August 1858.

34 Ibid.

35 Ibid.

36 'Steam Cultivators', *North British Agriculturalist*, 16 December 1863.

37 'Cultivation by Steam Power', *North British Agriculturalist*, 27 May 1863.

38 Ibid. We will get to a more detailed account on the competition between the Fowlers and Howards in Egypt in Chapter 7.

'A country predestined for the steam plough'

The *North British Agriculturalist* ran an extensive story on steam cultivators in mid-December 1863 showing improvements in machine ploughing since the summer of 1858. The price of a steam cultivator in England was between £500 and £1,000.[39] Farmers were hesitant, but those with bigger land and more capital were less reluctant to buy. In December also, a meeting was held by 'influential agriculturalists' in London to establish a General Steam Cultivation Company, to help small agriculturalists buy steam cultivators on credit.[40] The local British farmer foot-dragging seems justified according to some studies published by the Royal Agricultural Society after 1858. The studies relying on statistics from the field did not find a radical improvement through the steam plough over the horse in England: 'the accounts [of several farmers who used the steam cultivators for a year or more] show that the expenses of cultivating by steam power equal in amount that entailed by the employment of horses'.[41]

International demand, however, was high. At that moment, Fowler's steam cultivation system 'produced the greatest number of engines [. . .] and those are now in operation in almost every country, besides those in use [. . .] in most parts of Great Britain'.[42] The Australian colonies ordered three or four of Fowler's machines, as did the sugar plantations in the West Indies. The Cape of Good Hope, the East Indies, Austria, and Prussia – all had obtained some of the Fowler ploughs. However, 'Egypt is at present drawing more largely upon the resources of the [Fowler's] manufactory at Leeds than any other country'. The manufactory produced four machines per week, and there was an intention to double the production capacity, but 'by the latest accounts from Egypt, the demand for steam cultivators is so great that there will be some difficulty experienced in producing them in the various manufactories as fast as they are required'.[43]

Messrs Howard – the steam plough developers who were recommended the gold medal of the Royal Agricultural Society in 1858 – also

39 'Steam Cultivators', *North British Agriculturalist*, 16 December 1863.
40 Ibid.
41 Ibid.
42 Ibid.
43 Ibid.

sold portable engines, 300 sets, applicable to threshing and other farm power needs, probably since 1858. They shipped cultivators to France, Germany, Italy, Spain, India, and Portugal, and had received 'orders for several sets for Egypt'.[44] The *Cotton Supply Reporter*, on the occasion of an Egyptian order of eight steam cultivators from Messrs Howard around early December 1863, noted that a considerable percentage of the steam cultivators exported to Egypt were needed for 'preparing the land for the growth of cotton'.[45] The *North British Agriculturalist* added:

> The cultivation of cotton arising from the interruption of the supplies from the cotton producing states of North America, has tended to direct general attention to the steam cultivators, while the loss of working bullocks from the terrible epidemic which has swept away considerable proportion of the cattle, *has rendered it necessary to resort to some other means of cultivating the soil than by oxen.*[46]

The prime motive for Ismail buying British steam cultivators, ranging between £500 and £1,000 each, was certainly the cotton boom combined with the blow the cotton cultivation's prime movers sustained in 1863, as has been clearly demonstrated by this point.

What was it like to have this sudden and intensive influx of steam ploughs in late 1863? Fowler's representative in Egypt, engineer Max Eyth, spoke of a mayhem of machine parts in crates moving around to their setup destinations, in the Delta and across the valley. He also described an order from Ismail Pasha to Fowler to expand his workshop, to be able to produce and send more steam ploughs – 150 more to start with – and to bill the expansion in the Leeds workshops to the Viceroy's treasury. The success of the steam cultivator seems to have relied heavily on indirect but substantial Egyptian funding.

> Now there was a feverish exchange of telegrams between Cairo, London and Leeds. – 'When can Fowler's deliver 150 steam ploughs?' – Answer: 'Working to capacity, within one year.' – 'Not good enough.

44 Ibid.

45 'Steam Cultivation in Egypt', *Cotton Supply Reporter*, n.d., reproduced in *Bedfordshire Times and Independent*, 1 December 1863.

46 'Steam Cultivators', *North British Agriculturalist*, 16 December 1863. Emphasis added.

> Expect unloading Alexandria by spring [of 1864] 150 steam ploughs.' – A.: 'Impossible.' – The works at that time were barely big enough to turn out three steam ploughs per week [. . .] Ismail Pasha's next wire: 'Quote cost immediate factory expansion. Viceroy willing foot bill.' – You can imagine that Leeds made hay while the sun shone. And in addition, other factories in England and France as well were made to supply steam ploughs. The Alexandria warehouses, where goods destined for the vice-regal estates were unloaded, were crammed to the roof with boilers, wheels, drums, wire-rope and all sorts of chests and boxes. The second-rate hostelries of Cairo swarmed with newly qualified steam ploughmen, promoted in a hurry from anvil or share-plough, young hopefuls, fit for anything and nothing, since every steam plough must be manned by at least an expert pioneer of civilisation. Wagon loads of this assorted cargo were sent into the interior, just so that the next ship could unload. You cannot imagine in what condition they arrived at their destination, or rather anywhere but their destination. Ten boilers were lying on the banks of the Nile, and the machine to which they belonged was ten miles thither. Here was a little heap of wire-rope, but you had to travel another 20 hours to find the appropriate pulleys. In one place an Englishman who was to set up the machines squatted desolate and hungry, on a pile of French crates, and in another place the mate had taken to native liquor in his despair. Effendis and Katibs, invoking the help of Allah, rushed to and fro between Siut [Assiout] and Alexandria and compiled endless lists of items the names of which they did not even know. And yet, in the end, some of this apparatus was set in motion in Upper Egypt, the ploughs belched steam – civilisation and progress had made another step forward.[47]

We get a thorough account on the connection between the cotton boom, the dying oxen, and the radical expansion in adopting steam ploughing on the Nile, from M. George Ville, the superintendent of Napoleon III's experimental farm at Vincennes. He visited Egypt after the plague of 1863, and wrote an account of what he found, which was extensively quoted in the American *Hillsborough Recorder* years later.

47 Max Eyth, *Lebendige Kräfte: 7 Vorträge aus dem Gebiete der Technik [Living Forces: 7 Lectures from the Field of Technology]*, Berlin: Julius Springer, [1905] 1908, 21.

His report is thorough and indicative, and it comes shrouded in the romance known to have been expressed by early steam cultivation enthusiasts.[48] The *Recorder*'s anonymous reporter, writing in 1868, approached Egypt first by tying a certain steam cultivation revolution in the making there to the cotton boom caused by the American Civil War and the cattle plague of the mid-1860s:

> The war of the rebellion [the Civil War] in this country bravely changed affairs there [i.e. Egypt] as it did elsewhere. The high price of cotton, consequent upon the cutting off of the American supply, caused attention to be given to the inundated lands on the banks of the Nile, and cotton culture was rapidly introduced there. At first everything worked well, but the introduction of an additional supply of oxen from Europe was accompanied by the introduction of the rinderpest, which carried off not only the foreign importations, but the home supply as well.[49]

The report then moved on to M. Ville's testimony.

> At first confined to a few provinces, the pest gains ground and spreads like a vast conflagration. Oxen and buffalo die by hundreds, by thousands. The evil progresses constantly in extent and in intensity. The remains of the animals obstruct the canals and form floating islands in the Nile. And the price of cotton steadily advances. They console themselves for the loss of the animals – the gold of the cotton will pay for all! But the ploughing, the watering of the fields, the ginning of the cotton – without teams – how can it be done? In a few weeks the disorder became extreme. And then, that which neither good advances nor good example could have done[,] interest did.[50]

Remember the fences in England that blocked the path of the traction belt steam plough? And the inventors' objections that the Royal Agricultural Society was testing their machines on steep terrain? With

48 J. W. Y. Higgs, 'Review: God Speed the Plow, the Coming of Steam Cultivation to Great Britain', *The Agricultural History Review* 9:2, 1961, 136–7.

49 'The Steam Plow in Egypt', *Hillsborough Recorder*, 25 November 1868.

50 Ibid.

these in mind, M. Ville saw the steam-plough-friendly topography of the Nile valley:

> The land in Egypt is a limitless plain, as uniform as the surface of a lake. *It is a country predestined for the steam plough.* Without delay England offers her machines; she sends them by the cargo and on credit. They try, they hesitate, they feel their way; but the soil yields itself marvellously to the application of these implements. They become bolder; they familiarize themselves with the new machinery; they supplement it by the use of steam fire engine; and so well that last year [i.e. 1863–64[51]] I saw, in the valley of the Nile, machine shops fitted up as in Europe for the repairing and keeping in order of the most formidable apparatus that agriculture has ever employed. I saw ploughing done by steam engines of fifteen horse power, accomplishing an average of a hectare an hour.[52]

What the senior French agriculturalist did not factor in was that Egypt had been handling machines since the early nineteenth century, even if there were very few steam engines before the 1860s. Moreover, the presence of the Cairo–Alexandria railway since the 1850s, and the proliferation of steam cotton-ginning factories in the Delta witnessed at least since the late 1850s, probably provided technical expertise, foreign and local, and infrastructure for the rapid initiation of the many machine shops he saw.[53] He next turned his attention to the ceaseless toil of peasants, working alongside the modern machine cultivators:

> The men who work by day are at evening relieved by night gangs, and the plough continues its work by the light of lamps with

51 The *Hillsborough Recorder* dated Ville's account of steam cultivation in Egypt as follows: 'this description refers to 1865', and hence I deduced that in 1865, when he wrote this account, he was referring to 1864 or late 1863 when he said 'last year'. This guesswork is consistent with the timeline of the cattle murrain and other evidence we have here, and in the next chapter, on Egypt's purchases of the steam cultivators from England.

52 Ibid. Emphasis added.

53 We will see in the next chapter how three centres of steam machinery workshops might have been formed in the early 1860s, before the cattle murrain. In this we rely on evidence from Max Eyth, the chief engineer of Prince Halim's estates starting in 1862.

> reflectors. The plough is followed by rakes which form three ridges, eighty centimetres wide, on the top of which cotton is planted. Other machines are at work close by, converting the cotton wood [cotton shrubs' stalks] into faggots. Others dig canals; and all this machinery is kept in motion by steam, *which is produced by coal brought from Newcastle and Marseilles*. The machines were coarsely made and were sold at an excessive price. But all of that was details; the land was brought into cultivation and the cotton repaid everything with usury.[54]

With steam cultivation, the field hands worked more, not less, around the clock. So, how did the peasants perceive the machinery and this more intense workload? Did it go well?

> The Fellah made great eyes, struck with wonder and uneasy; he asked himself, not without apprehension, where all this was to end. But the profit was there real enough, represented by shining pieces of gold; so, little by little, the light has entered his soul. His hesitation has ceased; he commences to use steam fire engines in his villages, and associations are being formed for the purchase of steam ploughs.[55]

With this ends M. Ville's account on the decisive moment when steam engines proliferated in Egyptian agriculture, ushering in a comprehensive form of the new mode of energy and technology that Mehmed Ali had failed to bring in a few decades earlier. The testimony of M. Ville made a strong impression on the American reporter, pushing him to wonder why America was falling behind this underdeveloped country:

> It is a startling thought, that under the shadow of the Pyramids, which carry us back to the earliest barbarous civilization, two hundred sets of steam ploughing apparatus are now in use in a country that is no better adapted for the improvement than our western prairies, and among a people who are ignorant of the first principles of modern

54 'The Steam Plow in Egypt', *Hillsborough Recorder*, 25 November 1868. Emphasis added.

55 Ibid.

> mechanics, an almost universal familiarity with which constitutes our chief claim to material supremacy.[56]

The era of steam cultivation in America had to wait until the end of the Civil War in the mid-1860s, a story beyond the boundaries of this book. However, let us stop shortly at how the Fowler representative, Max Eyth, compared the situation when he moved to Louisiana from Egypt, suddenly in the late 1860s with the end of the Civil War, to set up more Fowler cultivators there.

> I sometimes examine myself to see if my growing dislike of America and the drab colour of the whole continent is because the stubborn Yankees are not buying steam-ploughs like my good Egyptians.[57]

Conclusion

In this chapter, we traced two timelines that intersected at the crossroads of the most substantial node of the energy transition in agricultural production, from animal to steam power, in the 1860s in Egypt. The scramble for expanded cotton production with the American Civil War, and Ismail's reacquisition of peasant land once he ascended the throne, brought the cattle murrain that brutally exterminated almost all the cattle that powered agrarian production. Meanwhile, the steam cultivator, following a much slower trajectory since the 1830s, started to be commercially viable at the end of the 1850s. Come the plague in 1863, and we hear of a sudden rush of steam cultivators to Egypt and its dying oxen. The Fowlers and Howards of England sold hundreds of engines to the country predestined for the steam plough, in M. Ville's words.

While the narrative given from M. Ville, and from other sources, points at a specific moment when steam-powered machinery abruptly flooded the Nile for agrarian production, there is more to investigate. We know, by now, about the contours of the technical advances in the production and commercial proliferation of the steam plough, and about the shape and colour of a transition on the Nile towards steam

56 Ibid.

57 Eyth, *Im Strom, Vol. I*, Chapter 116 (letter from New York, 28 October 1867).

cultivators, but there are questions that remain unanswered. First, how was the early transition to steam cultivators socially organised? Beyond the much-cited motives in sources from the period, that of oriental despotism and associated racially imagined laziness, stupidity, and irrationality of peasants, what were the motives of the sudden, as opposed to gradual and rational, importation of steam machinery? Who were the actors – Viceroy, richest landowners, and other peasant classes – who imported and deployed steam cultivators on their lands? The second set of questions is related to the state of the field in terms of agricultural steam machinery in Egypt after the 1863 transition, the fossil and non-fossil fuels for the machinery across the 1860s and 1870s, and the social impacts of this modern technology immediately in the wake of its deployment. The first set of questions will be tackled in the following chapter, while the second will be addressed in the rest of Part II.

7

The (Re)organisation of Power (1863–70)

> *'What do you think [Herr Eyth]? I want my castle here. Over there,' he pointed north, 'there are thousands of ducks and snipes, and pelicans and flamingos, and millions of fish [. . .] From here I see my cotton as far as you can see, and a dozen ginning mills, and fifty of your steam ploughs instead of the three puffing down there now! A good foundation [for the iron castle] should be provided immediately. Bringing in the castle and water and a little garden, that's all your business, Herr Eyth!'*
>
> Prince Halim, 1863, at Kafr Al-Sheikh

How was the early transition to steam cultivators organised between 1863 and 1870? Who were the actors who imported and deployed steam cultivators on their lands? In short, how was steam power organised, or rather reorganised, since it had already found its way to the Nile valley before the cattle plague did in 1863?

Answering these questions takes us into the realm of political power. Ismail Pasha reshuffled power arrangements including through the reorganisation of steam power, acting against European financial capital, and newly established local fossil capitalists from the Viceroyal family itself. The deployment of steam power across the later years of the 1860s came in combination with new attempts at monopolising steam-powered irrigation, using loans furnished by European financial capital.

This chapter accounts for the power dynamics – in both senses of the word – gleaned from the heat of the first substantial moment of the emergence of steam energy in Egypt (the 1860s) as it follows the steam engine's march towards becoming the dominant source of powering irrigation at least across the Delta by the early 1870s.

Steam Monopoly

There is abundant evidence on a steam-monopoly manoeuvre that the Viceroy attempted in association with importing hundreds of steam engines starting in 1863. Investigating this potential monopoly is integral to understanding the transition to steam technology, and how it was socially, economically, and politically processed and organised. This starts with accounting for engineer Lucovich's Agricultural Mechanical Company, possibly the first bearer of the word 'mechanical' in the name of any company in Egypt's history.

Antoine Lucovich, originally from Prčanj, Montenegro, served as, among other capacities, senior architect for Mehmed Ali at his palace in Alexandria, and then in the 1850s as the chief architect of the Austrian consulate in Egypt.[1] He requested and was granted a meeting with Ismail Pasha in May 1863.[2] It was a time of change. Over those weeks (late April to early June) the new Viceroy in power had started two major companies and a bank, in cooperation with prominent local and European businessmen and financiers mainly from Alexandria. Shortly before the meeting, Lucovich claimed, he published in the *Spetattore Egiziano* a detailed article with his plans for an agricultural mechanical company that was to revolutionise agriculture in Egypt by importing, selling, and installing steam-powered irrigation pumps. The plan was to sell and rent steam pumps to proprietors, and sell the water to those who did not wish to buy or rent the machinery, at a price lower than the cost

1 Udo Staf, *The History of the Lucovich Family: Private Ancestry Research*, presented at 'Egypt and Austria XII: The Current Research', Zagreb, Croatia (17–22 September 2018).

2 Lucovich stated that he was encouraged to meet Ismail after he saw him grant permission to two companies 'a few days earlier'. The official permission for one of these two companies is dated 2 May 1863. See *Nizamnamih* (bylaw) of the Egyptian Aziziya Company, Clause 1, in Amin Sami, *Taqwim Al-Nil 3.2 [Survey of the Nile Vol. 3.2]*, Cairo: Matba'at Dar Al Kutub, 1936, 585; Antoine Lucovich, *La Société Agricole et Industrielle d'Égypte*, Paris: Typographie de Gaittet, 1865, 11.

of irrigation by oxen and *saqiyas*.[3] Although he did not state it explicitly in his long testimony, this was the time of the first wave of the epizootic of the 1860s, and, understandably, when he talked about selling irrigation water, he meant that the peasants who were going to buy new oxen instead of the dead – or soon-to-die – were to have this option from his company. This was also the moment at which Ismail was about to start importing more oxen and steam cultivators in great numbers. As such, Lucovich either was interfering with Ismail's plans to import steam agricultural implements or did in fact plant the idea by explaining his plans to that effect.

The meeting was attended by the Austrian consul in Egypt. Lucovich knew that the presence of the representative of 'a great power' would make the Pasha more inclined to agree to his proposal. Ismail was indeed receptive of his plans; he even expressed admiration. However, 'he said, "for a company of this kind my authorisation is not necessary"'.[4] He reportedly mentioned many companies recently formed in Egypt without his authorisation, but Lucovich replied that the nature of his company, to be named Société agricole mécanique, meant that it needed to be allowed by local laws to operate inside the country, since its operations were to involve the setup of engines and machinery across the interior, overlapping and intervening mainly with the irrigation infrastructure. Apparently, to avoid saying a decisive no in the presence of the consul, the Viceroy told Lucovich that for such work across rural areas his company did not need Viceroyal authorisation. Instead, it needed to solicit approvals from local administrations across the country, adding that his own steam-powered irrigation pumps were always set up in ways that accommodated neighbouring landowners. Not relenting, Lucovich continued, through at least two more attempts at receiving the Pasha's authorisation, until the latter finally gave in, before ending the meeting: 'Eh! Well, I grant you [the authorisation], replied the Viceroy.'[5]

Lucovich applied for permission on 19 June 1863 and was handed a ministerial decision, he claimed, to that effect on 21 July.[6] Even

3 All information cited: Lucovich, *La Société*, 8–10.
4 Ibid., 13.
5 Ibid., 13–15.
6 Ibid., 16.

before the company permission came through, on 10 July he agreed with M. Edouard Dervieu – the prominent banker who handled the accounts of Ismail's mother and joined him in several companies – that the latter would recruit influential partners for the company. The morning after receiving the permission, on 22 July, Dervieu recruited major capitalists in Alexandria: Oppenheim, Ruyssenaers, Ross of Briggs and Co., Antoniadis, and others.[7] They developed the company's statutes, then sent them to Ismail for his consent. According to French consular correspondence, Ismail summoned his friend and favourite banker to Cairo and gave him orders to withdraw from the company.[8] Dervieu reported this back to Lucovich, who told him that he did not mind his departure, but the banker said that he would stay a partner, and that they should wait and see. He reportedly convinced Lucovich that he would visit the Viceroy again in Cairo to plead their case.[9]

Meanwhile, it seems that the Pasha issued orders to provincial *mudirs* on countering Lucovich's marketing activities outside Cairo and Alexandria. The latter, after hearing of the Pasha's disapproval of the company, moved on anyway to a marketing campaign in the Delta. Ragheb Pasha, Ismail's personal secretary, sent circulars, according to Lucovich, to all *mudirs* countering the advertisement Lucovich published to inform the public about the purposes and intentions of his mechanical irrigation company. When the engineer asked Ragheb for a reason, the latter said that the order authorising the formation of the company, from the Ministry of Foreign Affairs, did not conform fully with the language of the decision issued by another governmental agency to the said ministry about the authorisation. Something was lost in translation between Turkish and French, the two languages used in those two official documents. So, the Viceroyal authorisation was nullified for all practical purposes on inter-

7 David Landes, *Bankers and Pashas: International Finance and Economic Imperialism in Egypt*, London: Heinemann, 1958, 263.

8 French consular correspondences cited in ibid., 263. For an extensive archival investigation on Edouard Dervieu's many roles in the Egyptian economy, see Süleyman Uygun, 'Modern Misir'in Doğuşu (Yatirim, Kredi ve İflas Denkleminde) ve Bir Fransiz Aktör: Edouard Andre Dervieu [The Birth of Modern Egypt (in the Equation of Investment, Credit, and Bankruptcy) and a French Actor: Edouard Andre Dervieu]', *Studies of the Ottoman Domain* 9:17, 2019, 85–123.

9 Lucovich, *La Société*, 16–17.

ministerial miscommunication grounds, even if it was not officially withdrawn.[10]

Playing along, Lucovich sent the government – and the Viceroy – masked threats in the form of a letter to the *mudir* of Gharbiya, after the latter undermined his efforts to publicise the company's services to potential buyers from Tanta, in July 1863. In his letter, Lucovich said that His Highness would blame the *mudir* for this obstruction of the company's work since he earlier gave him his authorisation, but also this act would bring 'the severe criticism of all of Europe', since the *mudir* undermined Lucovich's 'commercial liberty'. Furthermore, the actions of the Gharbiya governor, in Lucovich's words, contradicted the Viceroy's rhetoric on progress and the march of civilisation.[11]

The following months witnessed negotiations and more interventions from Ismail. By July 1864, Lucovich managed to bring in a certain Colonel Comte de Kiss, a European capitalist without prior Egyptian presence or connections, to the prominent Alexandria capitalists.[12] Before the company's shares were floated in British stock markets in July 1864, however, the Ministry of Foreign Affairs, in documents declaring the government's approval of the company's statutes, changed the company's name to La Société agricole et industrielle d'Égypte, 'The Agricultural and Industrial Company of Egypt'. Lucovich was greatly concerned: 'to suppress the right to irrigate was to paralyse the forces of the company and dry up the most important source of income'.[13] This was the moment when the Austrian engineer started to question the good faith of his associates from Egypt – Dervieu, Oppenheim, and the others. The latter advised him, and advised the representative of the Comte de Kiss, to accept this change, claiming that they would discuss this problem later with the Viceroy, 'when the time is right'.[14] He suspected that Ismail was flooding his company with 'his auxiliaries' to undermine it gradually, especially when the directors insisted on doubling the paid value of the shares from 10 to 20 per cent – £4 paid by a shareholder for each £20 share. That made for a harder sell on the

10 Claims of Lucovich: ibid., 21.

11 Letter dated 5 August 1863, Lucovich to Fadel Pasha, *mudir* of Gharbiya. Fully reproduced in ibid., 18–20.

12 Landes, *Bankers*, 264.

13 Lucovich, *La Société*, 25.

14 Ibid., 26.

London market, even if Lucovich pushed for advertising the company in the British press as being primarily in the business of steam irrigation. The prospectus published in early July 1864 stated that the objective of the Egyptian Agricultural and Industrial Company was:

> to facilitate the means, and at the same time reduce the cost, of irrigation in Egypt, by the application of steam power in lieu of the primitive and wasteful mode at present in use, and also to encourage the economy of manual labour by the introduction of improved agricultural implements.[15]

The prospectus highlighted to potential shareholders the many good reasons for investing in such a company. It said that the cotton boom in Egypt had led to the appreciation of steam pumps by cotton-growing proprietors. It claimed that steam had the potential to irrigate 4 million more acres beside the 4 million then under irrigation, even if the population remained constant (5 million). The country had only recently lost circa £10 million as a result of the cattle murrain, 'a calamity which but for the introduction of steam power to replace the labour hitherto performed by oxen, would have paralysed the efforts of the cultivators, and deprived them of the means of availing themselves of the high profits to be derived from cotton cultivation'.[16] As for the buyers, they were to be 'proprietors and heads of villages' and the selling was to be 'either for cash, or payment taken by instalments. In which case, collateral security and interest will be required.' Finally, it announced that the most lucrative source of profit for the shareholders was to be the workshops established for repairing and maintaining the machinery. This maintenance business was estimated to 'secure a dividend equal, if not superior, to the most flourishing companies in existence'.[17]

One element in the prospectus that might indicate why Ismail wanted to neutralise this company and impose full governmental control over the setup of any new steam pumps across the country (more on that shortly) is the potential buyers of the engines: proprietors and heads of villages. The latter category should be understood as

15 'Prospectus', *Globe*, 8 July 1864.

16 Ibid.

17 Ibid.

indicating some sort of collective buying agreements coordinated by the heads of villages on behalf of the peasants. Otherwise, those figures would have naturally belonged to the category of proprietors. Imagine, in a post-murrain emergency situation – after the annihilation of animal irrigation power – a company established by rich European bankers and traders from Alexandria, with many European consuls backing them, coming to the interior with a governmental authorisation to provide irrigation power to the peasants. How hard would it have been for its representatives to convince the *sheikh al-balad* to talk the smaller landholders into buying shares in steam pumps, or the irrigation water itself through collective agreements, after they had lost their cattle? Too much was at stake for Ismail: irrigation fees to compete with taxes directly; offering credit in supplying the peasants with engines as substitutes of oxen and thus competing with tax collection indirectly; lost opportunities in selling the engines himself; and, most importantly, losing the power to intervene in distributing water that had always been in the hands of the state. Add to these risks the prospect of having the company make more money out of fixing and maintaining the engines through a monopoly over maintaining the steam machinery to be deployed – hence 'secur[ing] a dividend' that was to be more than 'most flourishing companies in existence' – and we can see Ismail cornered by a competitor who was aiming to remove enormous profits from his hands.

And the Pasha took action. In late July 1863, a year before the company's prospectus was advertised in London – actually the very week that Ismail met Lucovich with the Austrian consul – he issued two orders that 'have caused needless alarm among Europeans'.[18] The government banned the establishment of any 'steam factories without first obtaining the permission of the Government engineers, in order to guard against the danger of fire'.[19] The second edict 'forbids the erection of steam pumps on the canals', reportedly to ensure just distribution of irrigation water across the land.[20]

There is evidence that the governmental orders regarding steam were strictly followed. In the summer of 1864 an American citizen, one Mr

18 *Sun*, 6 August 1863.
19 Ibid.
20 Ibid.

Kindinco, started installing a steam engine (we can safely assume it was a Cornish steam pump) for agricultural work. It was claimed that the preparations at the engine's location destroyed part of a small canal supplying the adjacent village with water.[21] The government alerted the American consul, who intervened and stopped the work, though only temporarily. When Kindinco's labourers returned to the engine's site, the government 'sent a party of *cawasses* [law enforcement troops] to the spot, who forcibly ejected the workmen'.[22] The US consul demanded reparations for the insult of his flag. The government suggested sending one engineer to the site from each of the two parties to agree on a decision. The consul refused and gave the government twenty-four hours to yield or else the American flag was to be taken down at the consulate's building. The government ignored him, and the American flag was indeed taken down, after which the consul telegraphed a complaint to the Sublime Porte in Constantinople. Such was the extent to which Ismail went in enforcing the governmental oversight on deploying new steam engines.

It was around the time of the American consulate crisis that Lucovich was in Europe buying steam pumps. When he returned triumphantly with the engines in October 1864, he found that the board of directors were selling other engines for their own separate accounts and were preparing for a company reorganisation led by Dervieu. The latter thought that Lucovich's insistence on steam pumps would only bring the company – and his bank and other businesses in Egypt – the wrath of the Pasha. He was explicit in this during a meeting with the board of directors before 1865. They were convinced, and it was decided to move Lucovich to the more senior but less influential position of the comptroller general. Instead, Dervieu's brother-in-law was voted the manager, and another person close to the Ruyssenaers, and an employee of the Dutch consulate, was voted the commercial director. In March 1865, against Lucovich's vote, the board of directors agreed to buy the business of F. Basevi, a contractor and timber trader who reportedly had contracts with the Egyptian government for over £60,000. Upon the merger, Lucovich sold his shares in the company to its new manager, and the company's business was changed

21 'Egypt', *The Times*, 3 August 1864.
22 Ibid.

to public works constructions in Cairo and Alexandria.[23] Basevi's merger might have been just a detail in a larger understanding with Ismail by which he expressed his acceptance of giving the company free rein in such contracting for public works.[24] The company probably did not sell a single steam pump under the management of Lucovich or after his term.[25]

If the company established for importing and deploying the steam pump did not do such a thing, and was in fact blocked through an apparently elaborate plan, then who were the actors who brought the steam irrigation pump to Egypt in the early 1860s, and how was it deployed?

The Pasha, the Prince, and the *Pashmuhandis*

For the sake of making this complicated affair of organising steam power after the 1863 cattle murrain easier to follow, let us assume that it happened in trios. We have three prominent actors who were active in bringing in the steam cultivator. They were Ismail Pasha, his younger uncle Prince Halim, and *Pashmuhandis* (chief engineer) Max Eyth, the representative of Fowler of Leeds in Egypt, the manufacturer who more than any others, by far, produced and exported such engines. Another trio we should keep in mind (the focus of the next section) is the interplay between the mechanised cotton gin, the steam plough, and the portable steam pump. The latter two were different applications of the same steam engine: the portable steam cultivator/plough developed by Fowler, the Howards, and others.

Before delving into those two trios, we should appreciate first that the years 1863–5 witnessed an unprecedented leap in the quantity and value of steam engines and other steam-powered machinery exported from the United Kingdom to Egypt (Table 7.1). In the 1850s, Egypt's share of British global steam engines and other machinery exports ranged between less than 1 and 3.3 per cent (Table 7.1). During the first three

23 Landes, *Bankers*, 264–7.

24 Ibid., 266.

25 Roger Owen, *Cotton and the Egyptian Economy, 1820–1914*, Oxford: The Clarendon Press, 1969, 116.

Figure 7.1: (Left) Khedive Ismail, (centre) Prince Halim, (right) engineer Max Eyth
Sources: (Left) © Abdullah Frères (photographer)/Wien Museum; (centre) Sami, *Taqwim Vol. 3.2*; (right) © Museum Kirchheim unter Teck

years of the 1860s, British manufacturers sold in Egypt no more than 3.6 per cent of the total annual British engines/machinery exported. Then the Egyptian market's share rose sharply to 9.5 per cent in 1863, then 14.7 per cent and 13 per cent in 1864 and 1865 respectively, before falling again to a level higher than the early 1860s: 4.3 per cent. The steam engines exported from Britain to Egypt in 1864 were around one fifth of all steam engines exported from Britain to any destination across the world. Although the numbers are different, this leap did show in the same percentage (one fifth) in the estimates of the Manchester Chamber of Commerce. At the Chamber's annual meeting in January 1864, the chairman stated that British exports to Egypt in 1863, 'of steam engines, steam ploughs, cotton gins, packing presses, agricultural machinery, and implements of husbandry, were of the value of £91,000'. This compared to £80,000 for India, Singapore, and Ceylon combined, and £470,000 to all other cotton-producing countries.[26]

26 'The Cotton Supply', *Globe*, 26 January 1864. Concerning discrepancies between the Manchester Chamber of Commerce's estimates and those from the British official statistics: there are indeed different estimates depending on where you look, something that was highlighted by Roger Owen, but the general trend shown holds across the various numbers.

Table 7.1: Steam engines and other machinery exports to Egypt compared to steam engines and other machinery total exports of the UK, 1851–66

Year	Egypt's share of steam engines (£)	%	Egypt's share of other machin-ery (£)	%	Total value of all machin-ery imported by Egypt (£)	%	Global exports of steam engines (£)	Global exports of other machin-ery (£)	Global exports of all machin-ery (£)
1851	4,003	1%	3,066	0.4%	7,069	0.6%	403,637	764,974	1,168,611
1852	12,538	3.7%	15,965	1.7%	28,503	2.3%	338,222	913,138	1,251,360
1853	20,991	4.6%	14,300	0.9%	35,291	1.8%	458,376	1,527,160	1,985,536
1854	460	0.08%	24,980	1.8%	25,440	1.3%	566,768	1,364,092	1,930,860
1855	48,419	5.5%	25,656	1.9%	74,075	3.3%	883,370	1,359,796	2,243,166
1856	30,973	3.8%	19,768	1%	50,741	1.8%	819,067	1,897,386	2,716,453
1857	36,364	3.4%	70,732	2.5%	107,096	2.7%	1,069,249	2,814,420	3,883,669
1858	43,480	4%	37,028	1.5%	80,508	2.2%	1,097,278	2,502,074	3,599,352
1859	32,555	3.3%	64,525	2.3%	97,080	2.6%	973,340	2,757,961	3,731,301
1860	38,070	3%	69,457	2.7%	7,069	2.8%	1,238,333	2,599,488	3,837,821
1861	9,208	0.7%	62,362	2.1%	71,570	1.7%	1,258,164	2,955,506	4,213,670
1862	59,540	3.6%	74,854	3%	134,394	3.3%	1,624,876	2,467,797	4,092,673
1863	161,195	10.1%	252,521	9.1%	413,716	9.5%	1,595,638	2,772,976	4,368,614
1864	312,884	19.3%	402,071	12.4%	714,955	14.7%	1,617,117	3,231,475	4,848,592
1865	369,717	18.9%	311,974	9.5%	681,691	13%	1,958,533	3,264,100	5,222,633
1866	101,204	5.7%	105,428	3.5%	206,632	4.3%	1,760,612	2,998,479	4,759,091

Sources: Steam engines and other machinery exported to Egypt, Owen, *Cotton*, 100; steam engines and other machinery total exports of the UK, *Statistical Abstract for the United Kingdom from 1856 to 1870*, eighteenth number, London: George E. Eyre and William Spottiswoode, for Her Majesty's Stationery Office, 1871, 70–1; *Annual Statement of the Trade of the United Kingdom with Foreign Countries and British Possessions, for the Year 1855*, London: George E. Eyre and William Spottiswoode, for Her Majesty's Stationery Office, 1856, 16; idem, *1860*, 29.

Table 7.2: The value of UK machinery exports vs value of all UK exports to Egypt, 1861–6

Year	Exports of UK machinery to Egypt (steam engine and other machinery) (£)	Exports of all items from UK to Egypt (£)	%
1861	71,570	2,398,479	3%
1862	134,394	2,550,052	5.3%
1863	413,716	4,511,383	9.2%
1864	714,955	6,195,910	11.5%
1865	681,691	6,168,225	11%
1866	206,632	7,658,993	2.7%

Sources: Total value of machinery exported to Egypt calculated from: Owen, *Cotton*, 100; total exports from UK to Egypt: *Statistical Abstract, 1856 to 1870*, 16–17.

The suspect that caused this hike is the steam plough, as the murrain-induced energy famine of 1863–4 put pressure on finding alternative sources for powering ploughing and irrigation, including that of cotton. However, the earlier trend of expansion in steam-powered cotton ginning must have accounted for a substantial quantity of machinery imported between 1863 and 1865 as well.

When Ismail Pasha ordered his representatives abroad to buy cattle and other animals to replace the dead oxen, he did so with a view to resell most of the imports to other proprietors in Egypt. There are estimates that he did the same with most of the engines that followed, starting in late 1863.[27] This indicates that the transition was not only for his own estates, but also for big cotton farmers; for a while, he acted as the main importer of their engines and machinery. Let us briefly examine the evidence on the steaming of agriculture and cotton production by steam-powered irrigation.

The earliest information on record regarding when Ismail started thinking about the steam cultivator dates back to roughly one month after his meeting with Lucovich. On 1 June 1863 Ismail summoned Max Eyth and revealed his desire for an Egyptian threshing machine, asking Eyth to design it and produce a prototype. Eyth later commented, 'as His Highness's pangs of labour are very severe, I was summoned as a midwife, highly recommended by Halim Pasha'.[28] More importantly, Eyth gathered from the meeting that Ismail 'wanted to surrender to steam ploughing', after he witnessed the successful ploughing of Halim's land in Shubra with the steam cultivators the Prince bought from Eyth's employer over the previous couple of years. The engineer thought that this connection, born from this first meeting with the Viceroy, was of utmost importance to him and Fowler, undoubtedly thinking of the big purchase orders the Viceroy might have had in mind that summer.[29] From Eyth's letters, we can also gather that the first batch of steam

27 Landes estimates that the reselling of animals and machinery by Ismail was the reason why the 1863 budget did not show a big deficit (only 125,000 EGP in spite of the murrain and the flood). Owen agreed that this large-scale resale increased Ismail's income that year. Landes, *Bankers*, 191, footnote 2; Owen, *Cotton*, 117.

28 Max Eyth, *Im Strom unsrer Zeit. Erster Band. Lehrjahre* [In the Flow of Our Times, Vol. I: Apprenticeship Years], Projekt Gutenberg-DE, 1904 (letter 'Cairo, 21 June 1863').

29 Ibid.

ploughs Ismail ordered from Fowler came through in late October or early November 1863.[30] By 1 February 1864 or earlier, Ismail's orders of steam ploughs had led Fowler to treble the size of his workshops. Ismail did in fact pay for the expansion that made it possible to double their production rate of the steam cultivators, from four engines per week to eight.[31] At the funeral of John Fowler in December 1864, it was said that Ismail Pasha had in one order requested 120 steam ploughs:

> Some months ago while seated at breakfast he [i.e. Fowler] received a telegram with an order for 120 ploughs for the Pasha of Egypt. Their value is nearly £1,000 each – say an order for £120,000 worth of ploughs. These ploughs were to be used in cultivating cotton, and it was through the use of Fowler's steam plough that Egypt now in our time of need supplied us with cotton to the value of millions a year.[32]

And it was not only the Fowler workshops that sent steam ploughs over to Ismail. From Messrs Howard we hear that, in late 1864, they were contracted for fifty steam ploughs for Egypt. To put that in perspective, they also sold '*several* sets [. . .] in Spain, Portugal, Italy [for cotton culture], in France and Germany, *several* in India [for indigo farming]'.[33] At the annual meeting of the Manchester Cotton Supply Association, on 30 June 1864, the chairman said that the loss of the oxen in Egypt, which prevented her from fulfilling the expected cotton supply, was 'over-ruled for good, as on every side there were to be seen in that country machines worked by steam for pumping up water for irrigation; machines for cleaning cotton, also worked by steam; and steam ploughs almost exceeding in number those in use in England'.[34] He added, 'One large manufacturer of steam ploughs in the south of England had orders on hand from Egypt which would occupy him for some years.' A million bales of cotton were coming soon from Egypt, the chairman further claimed.

30 On 1 February 1864, Eyth suggested that Ismail got the first steam plough from Fowler three months prior. Ibid. (letter 'Shubra, 1 February 1864').

31 Ibid.; Max Eyth, *Das Agricultur-Maschinenwesen in Aegypten [Agricultural Machinery in Egypt]*, Stuttgart: J. B. Metzler'sche Buchhandlung, 1867, 60.

32 'Kendal Penny Readings', *Kendal Mercury*, 17 December 1864.

33 *The Farmers' Journal*, reproduced in 'Smithfield Club Cattle Show', *Bucks Herald*, 10 December 1864. Emphasis added.

34 'Cotton Supply Association', *Bolton Chronicle*, 2 July 1864.

There are other sources, however, that acknowledged the fact that agriculture was not yet fully revolutionised by the steam engines in mid-1864. The huge orders for agricultural engines from England started to make a difference after the spring of 1864. By late March 1864, according to the British press, animals and humans were still the pillar of agriculture, since 'hundreds [of steam ploughs] are on the way, but they will be late for this season'.[35]

Perhaps the great expectations of the Manchester Cotton Association cited previously were misdirected to the steam plough, given the incredibly large orders for that particular engine from a handful of manufacturers in England. At that moment, the momentum for importing cotton-ginning machinery and engines was still strong. That spring, a correspondent of the British press estimated that:

> While agriculture is the direction in which the most strenuous efforts are made, manufacturers are also more encouraged than they ever yet have been. Tall chimneys are shooting up into the air in many directions, and the noise of the steam engine or the factory hammer occasionally reminds the traveller of the busy hives of Western civilisation.[36]

The *Times* correspondent in Alexandria sang the same triumphant tone on the civilising role of Britain's steam power. In a long piece from March 1864 entitled 'The Progress of Egypt', which started with optimistic news on the Lucovich-led Société agricole, we are told that more cotton-ginning factories were coming, more mechanised flour mills, oil-seed-crushing mills, brick and tile factories, and so on. In short, 'Europe has finally understood the immense future of Egypt, and is eager to develop her yet budding resources'.[37] The cotton boom was finally starting to scar the skyline of the Delta with chimneys and smoke, and Englishmen were coming in numbers that matched those of engines:

> Not only are the cities of Alexandria and Cairo receiving so great an influx of inhabitants that, although whole quarters are rising on every

35 'Railway Works in Egypt', *Bicester Advertiser*, 5 March 1864.
36 'Foreign Policy: Egypt', *Bell's Weekly Messenger*, 11 April 1864.
37 'The Progress of Egypt', *Globe*, 28 March 1864.

> side, house room is still insufficient, and rents are always increasing; but the inland towns and villages are overrun, and factories with high chimneys and long lines of black smoke cut the clear sky of our flat landscape through the length and breadth of Lower Egypt [. . .] Englishmen, I am glad to say, are not behind in the race, *and their numbers must always increase in a corresponding ratio to the amount of machinery employed.*[38]

But if cotton gins and steam ploughs were driving the unprecedented hike in steam engine imports to Egypt and stirring this rhetoric on the civilisational role of Europe, who was bringing the steam irrigation pumps that Ismail barred Lucovich's company from supplying? There is evidence suggesting that steam ploughs were not used for the purposes of ploughing, but instead were likely customised on a great scale as irrigation pumps.

The Howard brothers in February 1864 accused Easton and Amos, consulting engineers in the capacity of arbitrators for the Royal Agricultural Society's exhibitions, of extending preferential treatment and better technical evaluations at the Society's competitions to engine manufacturers who bought and recommended their own agricultural implements to customers. One such implement produced by Easton and Amos was a pump, which we can understand from the correspondences between both parties as being sold by Easton and Amos in association with steam ploughs. 'Large number of pumps supplied for steam ploughing and other engines going to Egypt. One exhibitor, they are assured, has ordered some 50 pumps at about £40 or £50 each.'[39] By this the Howards were implying that Fowler was in the habit of buying pumping implements from Easton and Amos – for £40 or £50 a piece – as added implements for refitting steam ploughs as irrigation pumps, customised for Egypt. Easton and his partner Amos on 9 March 1864 sent a letter to the president of the Royal Agricultural Society replying to the Howard brothers' accusations. They explained that their workshops produced 'a large number of irrigation pumps we are supplying to accompany ploughing and other engines to

38 Ibid. Emphasis added.

39 Letter from Messrs Howard to Easton and Amos, 1 February 1864, partly reproduced in 'Messrs. J. and F. Howard and the Consulting Engineer of the Royal Agricultural Society', *Bedfordshire Mercury*, 27 February 1864.

Egypt'.[40] The letter further explained that they were the sole holders of a patent on mechanical pumping, and their pump was the only available such implement for agricultural steam engines manufactured for buyers in Egypt. This early in the process of purchasing steam ploughs, we must deduce then, the steam plough meant for the Egyptian importers, broadly, the mobile engines for agricultural purposes that gained recognition by the early 1870s or earlier as the locomobile.

From the start, the steam plough was deployed and used as a cheaper and more flexible alternative to the Cornish pump. Unlike the bigger beam engine, which irrigated hundreds of acres but was static in a building dedicated to it, the locomobile – able to irrigate around 100 acres – had the capacity of moving occasionally on wheels between different fields, as such offering through this mobility an irrigation capacity far exceeding its horsepower. It was invented to plough the land in England and Europe – where coal was cheap and could be burned in plenty to power its energy-intensive mobility – but was deployed in Egypt, more than any other country, for irrigation, something that Europe did not even need, as its agriculture was generally rain fed.

Why then did the British press not highlight this aspect of the steam cultivator/plough? Perhaps it would have made the steam plough look like a failure if its biggest customer turned out to be using it for other purposes (irrigation) that were redundant across many potential markets, such as Europe and America, for example. The manufacturers must have wanted to find other customers who would buy the plough for ploughing. For example, the American war was going to end, ushering in a huge potential market for steam ploughs. Ploughing also entailed the prospect of signing contracts for the maintenance of the engines and their locomotion parts, which were no doubt in need of repairs much more frequently compared to the immobile/fixed engine dedicated to irrigation only, which could be moved across fields only transiently, potentially pulled by oxen or camels. Its mechanised ploughing-related mobility was not necessary.

In the Egyptian market, the competition between Fowler and Howards in the spring of 1864 was fierce. Around April, Eyth learned of a representative of the Howards, Mr Delano, who moved to Cairo to

40 Letter reproduced in 'The Consulting Engineers to the Royal Agricultural Society', *Bell's Weekly Messenger*, 21 March 1864.

scout for potential buyers, and no sooner did he do so than he stumbled into Halim. Eyth claimed to have suggested to Prince Halim to organise a competition at Shubra between Fowler's and the Howards' ploughs. This idea of a plough race at Halim's Shubra steam technology laboratory might have been, in fact, Eyth's way of turning the tables on a demo the Howards' representative in Cairo requested for their ploughs for the Prince's eyes – to convince Halim to dismiss the Howards' engines once he saw Fowler's engine coming out superior. Whatever the case, Halim liked the idea and invited his nephew, Viceroy Ismail, to the show.[41] While Eyth claimed that Fowler's steam plough came victorious out of the competition held on 25 April 1864, the *Egyptian Times*, an Alexandria-based newspaper, claimed that the Howards' engine had beaten Fowler's in terms of speed and depth of ploughed soil.[42] Eyth was furious. He called the Howards' representative, Delano, 'a scoundrel', who corrupted the reporter and edited the piece of news in favourable terms to his employer. He challenged Delano to a pistols duel, then half-jokingly sent a friend of his to the hotel where Delano resided to spread the rumour that Eyth was a formidable shooter. Delano, in turn, sent back to Eyth a friend of German descent to talk him out of it. The duel never materialised after Delano, reportedly, apologised for interfering with the language of the press report on the race, but Halim ordered ten Howards engines nonetheless.[43]

The competition between Ismail (who started importing steam cultivators a few months earlier) and Halim (who generally invested in steam technology for years past) became explicit that same spring, before the former banished the latter to Istanbul for good in 1868 and confiscated/bought his property in Egypt. Steam cultivation was part of this story, and it is plausible that Halim's banishment was partly due to Ismail's concerns over – both steam and political – power.

Eyth's impression on Halim, his employer for several years and a lesser-known son of Mehmed Ali, is that he was a dreamer and avid believer in steam technology. In the late 1840s he had studied

41 Max Eyth, *Hinter Pflug und Schraubstock: Skizzen aus dem Taschenbuch eines Ingenieurs* [Behind Plow and Vice: Sketches from an Engineer's Pocketbook], Projekt Gutenberg-DE, 1910, section 'Eine Gegenmine'.

42 *Egyptian Times*, cited in ibid., section 'Nachspiel'.

43 Different segments of the story gathered from: Eyth, *Hinter Pflug*, several sections; Eyth, *Im Strom Vol. I* (letter dated 'Shubra, 18 May 1864').

engineering in Paris (École polytechnique), where 'it [was] said', according to Eyth, that the Prince had fought at the barricades in 1848 for the Republic, and was captured and then soon released, being a distinguished foreign student.[44] The same libertarian spirit apparently stayed with him. When, in 1868, he sent Ismail a public letter to protest at his nephew's harsh measures against him, he reminded him of the power of 'the public opinion', the only 'sovereign judge', since 'God is on high, and history is distant'.[45] At Shubra, among his entourage who wore Turkish attire, he was usually dressed as a European. He spoke to Eyth in French but the engineer was told by others to be wary, that the Prince understood English well, and 'his manners, apart from a tendency to sit on the floor anywhere, are entirely French-English'.[46] Standing with the engineer in his largely then-still-unimproved estates of Kafr Al-Sheikh, sometime before 1864, he brought up 'an iron castle', a magnificent hunting lodge in the middle of a marshland, that a French engineer had proposed to the Prince, and revealed to engineer Eyth his dreams about having fifty steam ploughs – from Fowler of Leeds – soon on that estate.[47]

Despite the fact that many other princes and rich land proprietors had, over the 1850s, deployed steam machinery on their estates, Halim was the first person in Egypt to contact John Fowler, in 1862 or earlier, once the steam cultivator became a practical invention, and ordered several mobile engines and an engineer to oversee them. By the summer of 1863, several reporters who covered the calamities of the cattle plague and Nile flood estimated that Ismail might follow his uncle Halim's example, and expected him to start importing the cultivators instead of the dead animals.[48] At that moment of rapid transition, no other person was named by reporters, besides Halim, as an already established steam cultivator in Egypt.

The Prince's relation to his nephew the Pasha – rarely explored in the literature – was tense, perhaps partly because of earlier family tensions and partly because of his lead in steam technology. He was one of two

44 Eyth, *Im Strom Vol. I* (letter dated 'Cairo, 17 March 1863').

45 'Halim Pasha', *Pall Mall Gazette*, 21 December 1868.

46 Ibid.

47 Eyth, *Hinter Pflug*, section 'Sackra'. Quoted in full at the opening paragraph for this chapter.

48 See cited news stories in Chapter 6.

princes on the train that sank in the Nile in 1858 and claimed the life of Ismail's older brother, the then-Crown Prince. Prince Ahmed drowned, but Halim was rescued by his personal guard, who happened to be a good swimmer. Ismail was supposed to board the same train but did not show up.[49] The perpetrator of this little-debated incident is still unknown, be it Ismail, Halim, Mustapha Fadil (Ismail's younger brother), Sa'ed Pasha, or simply a tragic accident.

The Pasha and the Prince might have enjoyed a year and a half of awkward peace after the former ascended the throne. Ismail appointed Halim the chairman of the Egyptian Commercial and Trading Company. It was originally conceived by Ismail for exploiting the then-still-unknown riches of Sudan, and for extending cotton cultivation to Upper Egypt, but, after it was pitched to European capitalists for financing, its operations base was changed to Egypt, and its earlier suggested name – 'Sudan Company' – was dropped.[50] Eventually, the company became a bridgehead for European financial capital into the Egyptian economy, involved in providing agrarian loans at 15 and 20 per cent annual interests.[51] A few months prior, in the summer of 1862 – a year before the cattle plague and half a year before Ismail became the Viceroy – Halim had caught European attention, including for his intentions to grow cotton in Upper Egypt by means of deploying steam pumps and ploughs on a large scale.[52] It is possible then that Ismail, once he took control in early 1863, might have wanted to co-opt Halim's planned cotton expansion in the south, flanking him with the honorary position of the president of the Sudan Company. Halim's agrarian business in

49 Two sources tell slightly different stories: Nubar Pasha, *Muzakirat Nubar Pasha* [Arabic translation by Jarou Robert Tabaqian of: Mémoires de Nubar Pacha], Cairo: Al-Shorouq, [1983] 2009, 269–71; Blanchard Jerrold, *Egypt Under Ismail Pacha: Being Some Chapters of Contemporary History*, London: Samuel Tinsley & Co., 1879, 20–3.

50 The company advertised first as 'Sudan Company' in July 1863 was said to bring cotton cultivation and cotton gins to Sudan (and Upper Egypt), and bring from it 'gums, elephant's teeth, bees-wax, ox-hides, gold dust', and cheap strong oxen. See 'The Alexandria Correspondent of the *Times*', *Bombay Gazette*, 2 July 1863.

51 Owen, *Cotton*, 115; Landes, *Bankers*, 151. It is very likely that Halim's role was honorary, a formality to extend Egyptian and Viceroyal colour to the company. Its board members included prominent European bankers in Egypt: Henry Oppenheim, Edouard Dervieu, and John Cater from the Bank of Egypt, as well as financiers from the Royal Ottoman Bank, and financial capitalists from London, Frankfurt, and Manchester, among others. See Landes, *Bankers*, 152, footnote 3.

52 *The Times*, reproduced in 'Egypt', *Kings County Chronicle*, 9 July 1862.

Upper Egypt reportedly extended also, in 1862, to 'competing closely with Ismail Pasha' in white sugar refining, a sector in which Ismail enjoyed singular excellence above all other local competitors, including his brother Mustapha, who inherited some of Ibrahim Pasha's sugar factories. Some sugar machinery that was heading that year to Halim's estates in Upper Egypt, probably at Muta'na, was described by the *Times* reporter as similar to those seen in the London Exhibition of 1862.[53] But more important, perhaps, is the fact that Halim was generally experimenting in steam machinery in agricultural production beyond economic calculations. He might have been the only person in Egypt in 1862 who invested in – alongside steam irrigation pumps and modern sugar-refining machinery – steam-powered reaping and thrashing machines and the Fowler-imported steam plough, at least two years before the invention caught Ismail's attention because of the cattle plague. *The Times* claimed in 1862 that, given Halim's efforts, 'it is not unlikely that a complete revolution may before long be established in the processes of Egyptian agriculture', a process through which agriculture might become 'more and more exclusively the pursuit of persons possessed of large capital'.[54]

Perhaps given this competition, Ismail appointed Halim in 1863 the head of the State Council, another honorary position that also helped keep the latter under the eyes of the former. Ismail must have felt Halim's great agrarian production advantages through steam cultivation once he started importing the steam ploughs in 1863. That year, the only mega-agrarian producer in the country with technical expertise and infrastructure necessary for deploying and maintaining new steam machinery at scale – beyond a few pumps or cotton-ginning establishments – was Prince Halim. Engineer Eyth oversaw the establishment of three workshops for maintaining the Prince's machinery, one at Shubra, another in Kafr Al-Sheikh, and the third in Muta'na in Upper Egypt.[55] To each of the Prince's estates, Eyth sent at least one English engineer trained at one of the three workshops, men who travelled to Egypt to work the engines even when many of them lacked relevant expertise.[56] For Ismail, who took

53 Ibid.

54 Ibid.

55 Eyth, *Das Agricultur-Maschinenwesen*, 55, 61; Eyth, *Im Strom Vol. I* (letter dated 'Shubra, 17 June 1865').

56 Eyth, *Das Agricultur-Maschinenwesen*, 61.

loans and imported huge amounts of steam cultivators in 1863 and 1864, the lack of comparable technical expertise brought such scenes that Eyth repeatedly described across his writings, while Halim's steam works were kept in nice order:

> In Alexandria [in November 1864], where I spent a few days with the intention of bringing order to the tangle of incoming machines, as far as they concern me, things are different. The acres of landing sites in the [Port's] 'Arsenal' are a mess of broken boxes, boilers, chimneys, bits of steel, cast, and wrought iron of all kinds. Trains loaded with machines stand for weeks on the tracks that disappear under mountains of equipment and cannot be transported due to a lack of locomotives. From the railway station through the city to the docks, the narrow streets are strewn with pieces of machinery: here a huge steam boiler, beneath which the marks of a collapsed wagon can still be discovered; there, a cotton press, half buried under dirt and dust, and apparently abandoned as irretrievable, there finally a broken box, grease guns, and so on.[57]

Ismail had indeed hired English engineers and technicians to accompany the imported machinery (including £200,000 worth of steam ploughs by late 1863[58]) but the sudden influx of experts – and many of them lacked relevant expertise anyway – for the incoming machines was not enough, producing scenes of complete chaos. Meanwhile, the Prince was quoted in the English press, with the advent of the cattle plague, boasting about a generation of local technicians and peasants who were taught steam technology on his lands:

> Halim Pasha [. . .] had found the intellect of the lads who have grown up since the introduction of the new mechanical appliances was greatly in advance of that of the men who had reached manhood under the former primitive system of cultivation, when the ox was the all in all to the fellahs, and when his mind had no stimulus and no cause for thought or inquiry.[59]

57 Eyth, *Im Strom Vol. I* (letter dated 'Shubra, 9 November 1864').
58 'The Viceroy of Egypt', *Dublin Evening Mail*, 3 November 1863.
59 'The Progress of Egypt', *The Globe*, 28 March 1864.

The Viceroy was quoted in the next line saying that the murrain was to 'confer a lasting benefit, by compelling the adoption of an improved system, which will civilise his people while enriching them', but he had nothing tangible to say comparable to his uncle with his already assembled steam expertise.[60]

Ismail moved against Halim. In late June 1864, he confiscated lands in the Delta, extending over several villages, that belonged to his cousin Tussun Pasha, the son of the late Sa'ed Pasha and Halim's nephew and foster son. In protest, Halim resigned from the State Council.[61] This land was in smaller plots, compared to Ismail's huge estates, potentially occupied by peasant households working for Tussun.[62] These might have been the same Delta swamps that Halim had protested at their Viceroyal confiscation from the villagers who turned them into productive land, or the 'rice villages' in the northern Delta that the Viceroy took over with the pretext that they were not productive enough.[63] He reportedly told Ismail that people might start saying the Viceroy is going to confiscate all villages across the country with the pretext of 'helping villagers'.[64]

Soon after, Ismail took 'all the workers at the steam ploughs in Shubra, machinists, and ploughmen, everything!', so said Halim to Eyth one morning.[65] When those technicians escaped at night and returned to their families on the Shubra estates, the Prince added, Ismail sent a military force that took them to the river port at Boulaq then on to his domains in Upper Egypt. When the engineer asked the Prince what he should do, the latter said, 'start over [. . .] If he takes away the second batch from you, start all over again [. . .] maybe it will do some good if our people are distributed across the country like this.'[66] The positive side of this affair, he concluded, was that it was going to rapidly expand steam technology learning at Eyth's 'Shubra University'.[67]

But it did not stop at technicians. Systemic depletion of Halim's production power might have been the next escalation. For example, in

60 Ibid.
61 Eyth, *Im Strom Vol. I* (letter dated 'Shubra, 21 June 1864').
62 Eyth, *Das Agricultur-Maschinenwesen*, 46.
63 Jerrold, *Egypt Under*, 88–91.
64 Ibid., 90.
65 Eyth, *Hinter Pflug*, section 'Im Sonnenschein'.
66 Ibid.
67 Ibid.

the spring of 1865, when Eyth gathered with great difficulty forty to sixty peasants for a job at Shubra, Ismail's men soon came and 'moved the whole herd into his stable'.[68] By January 1866, all the people who were not officially 'Halim's serfs' were being systematically and periodically taken over to the Khedival estates. These procedures might have been deliberate economic war tactics Ismail unleashed against several competitors beside Halim.[69] Technicians and engineers who learned the ropes on Halim's estates were being taken to build carts and run dredgers that dug the Suez Canal and many irrigation works sites. Meanwhile, the Prince seems to have lost interest in engineering altogether and took to fishing at some Delta wilderness with some of his guards.[70]

In November 1868, according to the supporters of Halim – himself turned in the 1870s into the most prominent dissident-in-exile – Ismail reportedly forged the signature of the Prince on a letter inciting forces in Syria to gather and invade Egypt, and he was also accused of aiding an assassination attempt on Ismail in Alexandria, in October.[71] Whatever the truth of this matter, on 3 November, Ismail sent his uncle a chilling ultimatum, to 'kindly leave the country within four days', to 'wherever Your Highness pleases'.[72] Almost two years later, according to a Viceroyal order to the minister of finance, Halim sent his personal physician back to Egypt to sign contracts on his behalf with Ismail, through which he sold all of his possessions, lands, palaces, and machinery. The same order included instructions for treating the Prince's family respectfully as they boarded a ship to join him in Istanbul, perhaps revealing that they were kept as hostages until the Prince signed the contracts.[73]

How far were Ismail's eyes fixed on political power or agrarian production power as he moved against his uncle? What motivated him more in his actions, getting rid of a strong political contender who had

68 Eyth, *Im Strom Vol. I* (letter dated 'Shubra, 16 March 1865').

69 'The Viceroy can at once ruin any of the native landholders by withdrawing the men from the estate and leaving the crop to rot': 'Notes from Egypt', *Newcastle Journal*, 12 July 1869, reproducing a report from a British correspondent from Alexandria to an unspecified newspaper.

70 Eyth, *Im Strom Vol. I* (letter dated 'Shubra, 5 January 1866').

71 'Halim Pasha', *Pall Mall Gazette*, 21 December 1868; 'The Sultan and the Viceroy of Egypt', *Glasgow Herald*, 17 July 1869; Jerrold, *Egypt Under*, 93.

72 Sami, *Taqwim Vol. 3.2*, 793.

73 Ibid., 867.

a claim to the throne after him or extending his control over the new production technology as it suddenly made its way with the oxen crisis of 1863 in the middle of the cotton boom? The answer is probably both, since both circles of power substantially intersected.

Since the late 1840s, those at the top of the newly formed landed class were members of the Viceroyal family, and they were consistently the first Egyptians to invest in fossil capital for agrarian production, beside and before the European fossil capitalists who invested in cotton-ginning factories in the Delta. While Halim's case is special, given how he seems to have been hooked to steam technology on (partially) ideological grounds, another contender for political power was not oblivious to steam technology. Prince Mustapha Fadil, Ismail's younger brother, who was likewise the target of the Viceroy's economic and political war, and who was banished to Istanbul two years before Halim, had substantial capital invested in adding steam-powered technology to his circles of agrarian wealth. The similarities between how Ismail outmanoeuvred him and Halim are uncanny. As with Halim, he started accusing him and his men of conspiracies. In April 1863, two top managers of his smaller *Daira* were accused of plotting against the Viceroy, arrested, and banished to Malta.[74] Like Halim, he had substantial estates that grew cotton and sugar cane across Egypt. In 1862, he and Ismail were the two top exporters of cotton, and his sugar factories and plantations in Middle and Upper Egypt inherited through their father Ibrahim Pasha, already mechanised to varying degrees, were substantial.[75] However, by 1869, they became the property of his brother Ismail, after the latter bought him out coercively in 1866, using yet another European loan, which was also used to buy his way through the Sublime Porte to turn succession rules from 'next oldest person in the family' to his own line.[76] The same agreement of 1866 included clauses giving Ismail access to a coal mine on Turkish lands, whose product turned out to be of inferior quality. Mustapha died prematurely in Istanbul, in 1875, reportedly in suspicious circumstances.[77]

74 'Egypt', *British Ensign* [newspaper], 8 April 1863.

75 'The Egyptian Cotton Crop', *Paisley Herald and Renfrewshire Advertiser*, 4 October 1862.

76 Edmond About, *Le Fellah: Souvenirs d'Egypte*, Paris: Librairie de L. Hachette et Cie, 1869, 349.

77 No title, *York Herald*, 8 December 1875.

Political and steam power, then, were integrated as parts of a whole. The reorganisation of one could not have happened without the other being involved.

The Gin, the Plough, and the Pump

We have seen how the Cornish pump and the modern sugar factory found their way to Egypt's agrarian economy from the late 1840s and through the 1850s. Then the steam-powered cotton gin made its grand entry with the cotton boom in the early 1860s. The cattle murrain, in 1863, brought the steam plough on a grand scale, the same mobile machine that was simultaneously easily fitted with a relatively cheap extension that turned it into a mobile irrigation engine. How did this trio – cotton gin, plough, mobile pump – fare by the mid-1860s and into the next decade? Which routes did this techno-energetic transition in agrarian production take?

There is evidence on two fronts. First, that the transition and importation of steam engines slumped by 1866 with the end of the American Civil War (in 1865) and the associated slump in cotton prices, before these imports rebounded in 1870. Second, that the steam plough was deserted, possibly for good, around the same time, in effect reorganising the setup of steam energy appliances in agriculture towards an increased share of steam in irrigation, mainly and at least in the Delta.

Table 7.3: Steam engines and other machinery exports to Egypt compared to global steam engines and other machinery exports of the UK, 1867–77

Year	Egypt's share of steam engines (£)	%	Egypt's share of other machinery (£)	%	Total value of all machinery imported by Egypt (£)	%	Global exports of steam engines (£)	Global exports of other machinery (£)	Global exports of all machinery (£)
1867	123,465	6%	33,080	1.1%	156,545	3.1%	2,026,072	2,942,436	4,968,508
1868	69,712	4%	40,608	1.3%	110,320	2.3%	1,724,783	3,004,699	4,729,482
1869	107,070	5.7%	100,024	3%	207,094	4%	1,851,779	3,267,143	5,118,922
1870	224,332	11.2%	91,317	2.7%	315,649	6%	1,997,523	3,295,750	5,293,273
1871	348,074	18.6%	296,976	7.6%	645,050	10.8%	2,064,004	3,902,037	5,966,041
1872	242,094	9.3%	164,887	2.9%	406,981	5%	2,594,996	5,606,116	8,201,112
1873	96,558	3.3%	110,845	1.5%	207,403	2%	2,927,617	7,092,312	10,019,929

1874	104,879	3.2%	67,995	1%	172,874	1.8%	3,255,685	6,535,229	9,790,914
1875	20,197	0.75%	77,615	1.2%	97,812	1.1%	2,631,333	6,427,314	9,058,647
1876	31,146	1.6%	55,710	1.1%	86,856	1.2%	1,939,117	5,271,309	7,210,426
1877	14,011	0.7%	43,363	0.9%	57,374	0.8%	2,018,864	4,704,004	6,722,868

Sources: *Annual Statement of the Trade of the United Kingdom with Foreign Countries and British Possessions, for the Year 1871*, London: George E. Eyre and William Spottiswoode, for Her Majesty's Stationery Office, 1872, 20, 193; idem, *1873*, 23, 195; idem, *1877*, 21, 191.

Suddenly and sharply, in 1866, Egyptian steam engine purchases from Britain dropped by more than 60 per cent, from a high of £370,000 the year before, to £100,000, and from accounting for around 20 per cent of global British exports of steam engines in 1864–5, to a 5.7 per cent share (Table 7.1; lowest point of 4 per cent was reached in 1868, Table 7.3).

In which ways can the cotton bust account for this change in conjunction with other political and socioecological factors? We know that the end of the American Civil War stymied the rapid upwards trend in Egyptian cotton production. From 2.5 million cantars in 1865, it dipped to 1.1 million cantars in 1866–7 (at 44 per cent of the higher volume) then ascended again to 2.5 million cantars by the mid-1870s.[78] But a dip around 1866–7 also could have been the result of the cattle murrain and associated sudden death in oxen used for ploughing across the Delta. While animal populations rebounded, it is plausible that smaller cotton growers had to divert their assets from growing more cotton to buying new oxen. Exports to Britain also fell sharply. Cotton exported from Egypt in 1866 stood at a lowest point of 67 per cent of its volume in 1865 (then rose steadily into the 1870s), while American cotton exports to Britain increased 380 per cent between these two years, and were up by 4,300 per cent from 1864.[79] So there was a strong and immediate effect of the return of the American cotton. Overall, it is legitimate to estimate that levels of cotton production, when they rebounded in the early 1870s, stabilised at Egypt's global market share in the mid-1860s, not more, given the American market's return.

But if the cotton production and associated exports slowed down transiently then rebounded to 1865 figures throughout the next decade,

78 Owen, *Cotton*, Tables 7 and 14, [pp.] 90, 123.

79 *Statistical Abstract for the United Kingdom from 1856 to 1870*, eighteenth number, London: George E. Eyre and William Spottiswoode, for Her Majesty's Stationery Office, 1871, 58.

why did British steam engine and machinery exports to Egypt maintain a steady but lower level throughout the late 1860s? The sharp increase to near 20 per cent (£348,000 worth of steam engines by 1871) can be partially accounted for through Khedive Ismail's sudden turn to heavy investments in sugar refining in 1870–4. Approximately twenty modern factories were imported from England and France, using the 'sugar loan' of 1870. But why maintain an average of £100,000 steam engine imports between 1866 and 1869 (Tables 7.1, 7.3; Figure 7.2)?

The first suspect is the dredging of the Suez Canal, radically steamed and mechanised between 1864 and 1869, as machines replaced corvée around the mid-1860s. However, evidence suggests that the great assembly of dredgers and other steam machinery bought for the job was overwhelmingly of French origin. Engineers Paul Borel and Alexandre Lavalley were the main contractors for providing the Suez Isthmus mechanical dredging works with machinery between 1864 and 1869. With them came the Dussaud brothers of Paris and other fellow francophone industrious citizens, and it might have been the case that all of the canal dredging and digging machinery was brought from France, since it was returned there periodically for maintenance and redesigning on a needs basis. No British fingerprints here.[80]

Figure 7.2: Egypt's steam engine and machinery imports from the United Kingdom, 1851–77

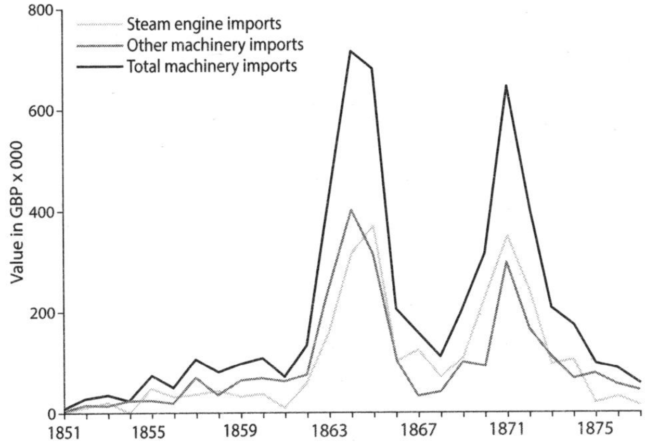

Source: Tables 7.1 and 7.3

80 Zachary Karabell, *Parting the Desert: The Creation of the Suez Canal*, New York: Random House, 2004, Chapter 16.

It is likely then, absent other major suspects, that the British import of steam engines of 1866–9 was mostly the same steam cultivators brought for the agrarian economy since the year of the murrain (1863). How did agrarian production applications of steam technology in Egypt unfold with and immediately after the American Civil War?

In early 1864, *Pashmuhandis* Eyth reported from Halim's Shubra estates that steam ploughs were working ceaselessly, across the night, for the first time in history, he said. He realised that he and Halim 'have created an agricultural machine economy here that, to [his] knowledge, does not exist anywhere else in the world'.[81] In 1865, however, even with the cotton crisis setting in, with seasonal agricultural workers getting laid off all of a sudden, and steam ploughs sitting idle for the first time in two years, 'there was no stopping of steam technology taking hold of the field'.[82] With the cattle population rebounding considerably that year, Eyth added, many of the steam ploughs bought for ploughing – not irrigation – were refitted with irrigation gear, and turned into pumps.[83] Simultaneously, with the looming end of the cattle murrain (making it less risky to import new animals to do the ploughing), and lower cotton prices on the market (making it less economic to invest in expensive steam plough maintenance and periodical replacement of parts needed for its ceaseless mobility over heavy alluvial black soil), the oxen returned to the fields. The trend that held, however, was in all likelihood the importation of more steam ploughs retrofitted as irrigation pumps. The steam plough appears, in hindsight, then, as having been nothing more than an emergency fix for the earlier-booming cotton-led agrarian economy that brought with it the cattle murrain.

Anecdotal evidence supports Eyth's evaluation of this post-murrain phase of steam cultivation. Right after the cattle murrain, in the spring of 1864, a British tourist saw, at an 'oddly named [train] station very close to Alexandria', two steam ploughs at work in a huge cotton plantation belonging to the Viceroy.[84] A few days later in Cairo, he saw at his hotel six young Englishmen who were visiting to set up and superintend 'steam

81 Eyth, *Im Strom Vol. I* (letter dated 'Shubra, 1 February 1864').

82 Eyth, *Das Agricultur-Maschinenwesen*, 71.

83 Ibid.

84 Archibald Black, *A Hundred Days in the East*, London: John F. Shaw & Co., 1865, 30.

ploughs and cotton-cleaning (ginning) machines for the Government and private capitalists.'[85]

After 1865, other travellers had noticed the proliferation of ginning and irrigation steam machinery. The mechanisation of cotton ginning was even extended to corn threshing: 'some of the great men have lately [from the perspective of 1869] introduced threshing-machines on their estates.'[86] A British canoer visiting Egypt in 1868, with weeks spent on the Nile, lakes, and irrigation canals, spotted and heard the 'puff puff[ing]' of high-pressure engines used for pumping water in the Delta. He noticed that the capital city of Daqahliya was 'immersed in cotton', which 'bursts forth on all sides'. 'At Mansoura the cotton-gins for cleaning the stuff and separating the seed are worked by steam; and ceaseless sighs of the engine are heard.'[87] Meanwhile, there were no mentions of steam ploughing.

There are structural economic and ecological reasons that might tilt the balance greatly to assuming that the greater chunk of steam engines imported from England across the late 1860s was steam pumps. Perennial irrigation, radically expanded and systematised across the Delta since the 1820s, was increasingly taxing for Ismail and landholders big and small by the 1860s, especially with the cotton boom. Summer canals that fed the cotton and other summer crops – including sugar cane – tended to silt and clog because of their greater depth accompanied by weaker slopping, and as such required increasing volumes of labour for seasonal clearing as they multiplied in association with increasing cash-cropping. After the completion of the Delta barrages in 1861, the purpose of which was to raise the water level in the Delta summer canals to avoid silting and clogging (and as such decrease the labour needs for clearing), it suffered cracks and other structural problems the first year of its operation, and, by 1867, its Rosetta branch was further compromised, requiring an estimated five-year repairs programme budgeted at 1 million EGP. Instead of the hefty repairs, M. Linant, the trusted engineer who had served in this role since the reign

85 Ibid., 32.

86 M. L. Whately, *Among the Huts in Egypt*, London: Seeley Jackson & Halliday, 1871, 3.

87 J. MacGregor, *The Rob Roy on the Jordan, Nile, Red Sea and Gennesareth: A Canoe Cruise in Palestine and Egypt, and the Waters of Damascus*, London: John Murray, 1870, 50, 58.

of Mehmed Ali, recommended supporting summer irrigation by importing steam pumps.[88] This potentially justifies the £100,000 average British steam engines bill across 1867–9, even after cotton prices slumped and its production decelerated for a few years.

This fix to perennial irrigation problems for cash crops, which partly shifted the maintenance costs from the state budget to the agrarian producers, seems to have carried on into the 1870s. By the early to mid-1870s (according to a survey published in 1875 by the irrigation engineers of the Ministry of Public Works and compiled by the minister Ali Mubarak), the provinces of the Delta had among them 1,320 steam pumps, with the combined capacity of 14,193 hp.[89] The survey included two categories of machines, 'fixed' and 'mobile', the first described as *Thabit* (Arabic for fixed), and the second as *Comobil* (short for locomobile). The latter category formed the bulk of the steam-pumping power of the Delta by then.

Let us look at a few examples. The Ja'fariya irrigation canal in Munufiya had fifteen steam pumps installed; only three were fixed (30 hp combined) and twelve were of the mobile type (114 hp). At Nabarouh canal and Beyala all twenty pumps were locomobiles, with a combined power of 183 hp.[90] On Abi Al-Akhdar canal in Sharqiya, twelve pumps were installed, one fixed (50 hp) and the rest locomobiles (114 hp).[91] Al-Mansouriya canal, dug in the early nineteenth century by Mehmed Ali, and one of the canals on which the British canoer heard steam pumps puff-puffing, had by 1875 around fifty-two steam pumps, eleven fixed (158 hp combined) and the rest locomobiles (292 hp).[92] On the Behira province's major canal of Al-Khatatba, only seven steam pumps

88 Aaron Jakes, *Egypt's Occupation: Colonial Economism and the Crises of Capitalism*, Stanford: Stanford University Press, 2020; R. H. Brown, *History of the Barrage at the Head of the Delta of Egypt*, Cairo: F. Deimer Publisher, 1896, 18–21.

89 Ghislaine Alleaume, 'An Industrial Revolution in Agriculture? Some Observations on the Evolution of Rural Egypt in the Nineteenth Century', in *Agriculture in Egypt: From Pharaonic to Modern Times*, edited by Alan Bowman and Eugene Rogan, Proceedings of the British Academy, Vol. 96, 1999, 339–40.

90 Ali Mubarak, *Al-Khitat Al-Tawfiqiya Al-Jadida Limisr Al-Qahira wa Mudunaha Wa Biladha Al-Qadima Al-Shahira 19* [The New Tawfiqi Encyclopaedia of Egypt's Cairo, and Its Other Old Cities, Towns, and Villages Vol. XIX], Cairo: Al-Matba'a Al-Kubra Al-Amiriya, 1889, 26.

91 Ibid., 52.

92 Ibid., 68–9.

were fixed from among eighty (with 82 combined hp). The locomobiles (seventy-three) accounted for 656 hp.[93]

Table 7.4: Average irrigation water (cubic metres of water/day) consumed in the summer season of 1290 H (1873) across the Delta

Province	Direct from Nile and canals (m^3)	Elevated by steam pumps (m^3)	Elevated by oxen-led *saqiyas* (m^3)	% steam pumps to *saqiyas*	Total (m^3)
Qalioubiya	370,000	300,000	250,000	120%	920,000
Daqahliya	1,100,000	400,000	300,000	133%	1,800,000
Munufiya	600,000	300,000	100,000	300%	1,000,000
Gharbiya	1,100,000	400,000	200,000	200%	1,700,000
Behira	1,200,000	400,000	300,000	133%	1,900,000
Total	4,370,000 (60%)	1,800,000 (24.5%)	1,150,000 (15.5%)	156%	7,320,000

Source: Mubarak, *Nukhbat*. Sharqiya is excluded since its sources of irrigation were not detailed like the other provinces'.

To put the 14,193 hp steam-pumping capacity in perspective, it is noteworthy that this steam irrigation power was, in all likelihood, greater than oxen-led irrigation, even by 1873. Before or in combination with works related to this survey, Ali Mubarak estimated that steam pumps across the Delta processed 156 per cent of water elevated to lands by oxen-led *saqiyas* any given day of the summer season (with the highest ratio of steam pumps to *saqiyas* in Munufiya, at 300 per cent; Table 7.4).[94] We should also factor in that the mobility of the locomobile pump – which formed the bulk of steam irrigation power by the early 1870s – made it possible to use the engine across several plots of land, as such extending the actual horsepower beyond these numbers. It must have been easy to hook the relatively small irrigation engine on wheels to a pair of oxen and move it between fields, in case such an engine had, with longer times staying idle, rusty ploughing/mobility parts and/or in the absence of experienced technicians and drivers to make this happen, or in case it would have consumed large amounts of expensive coal for the trip. We do not know whether Ali Mubarak factored this in when he came up with the numbers on steam pumps to oxen-led *saqiyas* covered previously, but, in his 1873 book cited (*Nukhbat Al-fikr*), he was

93 Ibid., 76–8.

94 Ali Mubarak, *Nukhbat Al-fikr fi Tadbir Nil Misr* [Selected Thoughts on the Management of the Nile], Cairo: Wadi Al-Nil Press, 1880, 89–90.

advocating the use of steam power in irrigation as well as canal dredging much more beyond the proliferation levels already reached.[95] For example, he provided evidence-based estimates that installing 5,300 hp more of steam irrigation power (centrally planned by the state and strategically deployed) in four of the five provinces of the Delta – for no more than 320,000 EGP – would increase cotton yield by one cantar/acre across the Delta.[96]

One last note on the steaming of irrigation in the Delta across the early 1870s is that small landholders might have transitioned to steam pumps partially as a form of paying an indirect tax. With Ismail choosing to ignore maintaining the Delta barrages in 1867 over its hefty bill, and leaving summer irrigation to steam pumps – while spending more money on digging new summer canals, most notably the extensive Ibrahimiya canal of Middle and Upper Egypt, completed in 1873 – roughly 33 per cent of the Delta's total irrigation horsepower in 1875 was owned by smaller landholders.[97] Many of the steam engines of 6–10 hp each (mostly locomobiles) in Ali Mubarak's inventory belonged to men without any honorary titles (not even Effendi), and many of those belonged to two or more owners.[98] A significant proportion of these might have been bought through village cooperatives, through which several small landholders co-funded a single mobile pump and took turns in using it for irrigation. At least since 1865, peasants were reportedly using these collective arrangements to replace their dead oxen with steam ploughs.[99] Over earlier centuries, it was common to collaborate in buying a single ox or buffalo.[100] There is no reason why they might not

95 Alleaume, 'An Industrial', 339–41; Mubarak, *Nukhbat.*

96 Mubarak, *Nukhbat*, 123–5.

97 4,708 hp from 14,193 hp. Calculated from: Alleaume, 'An Industrial', Table 15.2.

98 Mubarak, *Khitat Vol. XIX*. Examples of co-owned pumps: 'Ali Salih and Partners, 1 pump, 10 hp; Al-Sukari and Partners, 1 pump, 10 hp', 42, and 'Mustapha Abi Harga and his cousin Mohamed, 1 pump, 6 hp', 'Mohamed Al-Manzalawi and his partner Abdulsalam Abdulrahman, 1 pump, 8 hp', 40.

99 'The Steam Plow in Egypt', *Hillsborough Recorder*, 25 November 1868: '[The peasants] commence[d] to use steam fire engines in their villages, and associations are being formed for the purchase of steam ploughs.'

100 Alan Mikhail, *Under Osman's Tree: The Ottoman Empire, Egypt, and Environmental History*, Chicago and London: University of Chicago Press, 2017, 116: 'The *jāmūsa* [water buffalo] was so expensive an animal that peasants often had to pool their resources to be able to purchase one. In these cases, peasants bought shares in a *jāmūsa* allowing them partial rights to the animal's labor, milk, and other productive capacities.'

have deployed and upscaled these arrangements – by the late 1860s – to secure their summer crops against the deteriorated canals and barrages, by purchasing locomobiles.

Middle and Upper Egypt's steam irrigation power around the same time is harder to reconstruct. It was not covered by Ali Mubarak's survey, and in the absence of substantial plantations of cotton south of the Delta, we might expect that adopting steam power in irrigation above Cairo was far behind the situation downstream. A rare piece of information from 1870, however, offering a detailed inventory of steam pumps on Ismail's Daira Saniya plantations south of Cairo, suggests otherwise. As part of efforts to support the shares of the 'sugar loan' of 1870, which the Viceroy raised through European bankers for £7 million in order to import over a dozen sugar factories from France and England, his minister of finance, Ismail Saddik Pasha, privileged the *Railway News* with an exclusive inventory of some of the *Daira* assets, in September 1870. This inventory included specifically the assets tied as security to the loan, exclusively from Bani-Suef, Minya, and Assiout.[101] We learn from the minister's list that Minya province alone, even before the upgrading of its sugar factories, possessed fourteen 'large pumping establishments' at a combined capacity of 1,350 hp. Assiout's *Daira* possessions included nine steam pumps of the combined capacity of 667 hp. In Biba, Bani-Suef, eight steam pumps had the combined capacity of 630 hp. In all, across the Daira Saniya in parts of Middle and northern Upper Egypt alone, in 1870, we have estimates of around 2,650 hp steam-pumping capacity, set up across 200,000 acres, the part of the *Daira* in Middle and Upper Egypt that was directly cultivated by Ismail's administrators, or particularly across 157,000 acres, the land inventoried by the minister of finance in the exclusive *Railway News* list.[102] The steam engines at the then existing sugar factories – for crushing cane, heating juice, and refining sugar – were not included in the inventory.

If the Middle Egypt *Daira* estates alone in 1870 had 20 per cent of the irrigation steam horsepower of the Delta as calculated in 1875

101 'The Khedive Loan', *Railway News*, 17 September 1870. The inventory's focus on assuring the loan's bondholders (and supporting its floating value in the market) was stressed towards the end of the inventory. This might suggest that the hp of steam irrigation capacity provided was exaggerated.

102 Ibid. J. C. McCoan, *Egypt As It Is*, New York: Henry Holt and Company, 1877, 147. 'The Khedive Loan', *Railway News*, 17 September 1870.

(over 14,000 hp), then we might assume that richer landholders in Middle and Upper Egypt possessed substantial steam irrigation power as much as their counterparts did in the Delta. The richer owners of steam pumps in Mubarak's survey of 1875 outside the Viceroyal house – those with titles of Pasha, Bey, Effendi, Khawaja, Agha – had among them 5,756 hp (40 per cent of Delta irrigation mechanical horsepower).[103] In short, the adoption of steam power for irrigation in Middle and Upper Egypt might have been closely following the trend in the Delta. But even if we drop this data on Middle and Upper Egypt, assuming it is less credible than the Ali Mubarak survey – that is, that the hp listed was exaggerated to support the value of the floated loan bonds – it is still legitimate to assume that steam-powered irrigation, at least in the Delta, was taking over animal power, with an upwards trajectory, and that the steam plough had a great share in this as it was retrofitted into irrigation roles.

Conclusion: The Reorganisation of Power

Let us recap the career of the steam technology in agriculture so far. With the cotton boom, first came the cotton-ginning factories in the early 1860s, across the later years of Sa'ed's reign. The next phase was the arrival of the steam cultivator (both a plough and pump) as an emergency measure with the cattle murrain. With it, technological innovation and expansion at Fowler's workshop and others – the whole economy of the new technology expanded, at least in England, an expansion partially financed by Khedival money – helped make the steam irrigation option more economic in the second half of the 1860s, when the oxen returned and cotton prices slumped. The bust made steam ploughing uneconomic, given its higher costs of maintenance relative to lower cotton prices and the much greater amount of coal it required to enable mobility. Across the last five years of the decade, we see more of only two of the trio: the cotton gins and pumps. Mentions of steam ploughs become almost non-existent after 1865. By the late 1870s, not a single account of machinery in agriculture in Egypt that came to my attention mentioned steam ploughing.

103 Data calculated from: Alleaume, 'An Industrial', Table 15.2, 340.

The steam cultivator (as plough) became extinct on the Nile after a brief and sudden emergency deployment to replace the dead cattle between 1863 and 1865, but was reborn as the locomobile, which led the systematic proliferation of steam power in Delta irrigation until it surpassed, by far, the animal/human irrigation energy regime by the early 1870s.

Since day one, steaming the agrarian economy, or steam cultivation, involved an entanglement of economic and political contentions featuring Ismail Pasha, some members of the Viceroyal family, and European capitalists in various capacities. On the one hand, after the European cotton gin factory owners across the Delta and all those who invested financial capital in extending loans to cultivators came the fossil capitalists who developed the steam cultivator, and competed over greater market shares in Egypt: Fowler and the Howards for example, and Easton and Amos as sellers of pumping gear that retrofitted the plough as locomobile. On the other hand, there were the financial capitalists and entrepreneurs, for example engineer Lucovich and some of his associates, who craved all sorts of profit through the steam pump and monopolising irrigation by controlling its mechanisation. The struggle over having the upper hand in the deployment of steam power, ultimately for cotton production during the boom, was both a prime motive for Ismail's loans from Europe and a strong shaper of his palace ploys, against his uncle Halim, for example.

Falling back to the general contours of this complicated story, let us ask again the question we set out with in this chapter: how was power (re)organised around steam power across the 1860s?

Ismail Pasha's interventions in this reorganisation seem to have been planned and conducted in a way that sought to keep him on top of the recently formed (since the mid-1840s) class of landed proprietors. He became the sole big importer of agricultural engines and redistributor of these engines, and he moved against Halim, afterwards confiscating his estates to claim the most superior steam expertise in the country thus far. This process was enabled by loans from Europe. The two loans of 1864 and 1865 were – either directly through buying or indirectly by supporting a depleted budget – the source of reorganising the means of agrarian production, at least in the Delta, in a way that assured the head of the state the prominence he sought as he came to power, far beyond the limits of the state budget. The murrain was the cotton boom's 'natural' disaster, society/nature's combination of the cotton boom. Structurally speaking, it was the result of expanded mono-cropping of cotton with the cotton boom of the early 1860s. This expansion was accompanied by the same tendency witnessed in the 1840s murrain when land was first redistributed from peasants to big landholders. In other words, the loans financed the energy transition orchestrated by the Viceroy – which was itself accelerated as the result of the cattle plague: cotton boom's impact on nature – and were necessary in the equations of the rearrangement of political power in a way that kept the Viceroy on top of the landholders' class and financial capitalists of Europe.

In hindsight – given the state bankruptcy with loan defaulting and ensuing encroachment of European capital and military power over Egypt across the late 1870s and then the decisive moment of British occupation in 1882 – those very same loans served as bridgeheads for unprecedented levels of European financial capital into Egypt. While these loans and others were used to maintain prevailing social and political formations by their taker, they were spent in ways that led to a phase of horizontal penetration of capitalist production. The following is a non-conclusive list of this horizontal penetration: the establishment of modern factories – cotton gins and others in the Delta across the 1860s, then sugar factories starting 1870 – that gradually brought the modern factory system of paid wage labour; extension of railways across the country; sustaining and increasing mono-cropping in cotton and sugar, thus strengthening ties to the global economy; monocrops-associated deepening of monetisation and credit; proletarianisation of more peasants as big landholders consolidated more land with the further commodification of land. In short: the consolidation of soft and hard infrastructures of an agrarian capitalist formation were being deployed using European financial capital through the loans.

While it was happening, the reorganisation of energy and power – power in both senses of the word – had the effect of boosting the prevailing pre-capitalist mode of labour exploitation by giving it a lifeline that extended it to building the infrastructure necessary for making Egypt a more efficient monocrop producer. At the same time, the loans opened up, through their inherent longer-term tendencies, the progression of the integration of Egypt into the Capitalocene that originated from western Europe. The bankruptcy of 1876, ensuing dual British–French financial control procedures, and subsequent occupation in 1882 can be understood then as foundational for the vertical concentration of capital phase.[104] With it, European financial capital penetrated, side-lined local intermediaries, and exploited the local labour force and natural resources with its own hands.

But it was not a thoroughly planned and executed capitalist plot. Again, capitalists seek simply to increase surplus value, but when they do, certain tendencies seem to start operating, which we can only try to analyse and interpret following the history, not the laws per se. Moreover, it was an uneven transition, geographically – across the Delta and Middle/Upper Egypt – and non-linear, at least in terms of the nature of the fossil capital deployed. What we have on both counts is, first, that the transition associated with the cotton boom in the Delta – and accompanying socioecological, political, and economic processes of the early 1860s at large – manifested itself in northern Upper Egypt in the form of a wide-scale popular revolt (the next chapter). Second, that the deployment of steam-powered technology in Egypt did not necessarily use fossil fuels – coal in our case – across the board. There is overwhelming evidence that Khedive Ismail's agrarian production, and that of other actors in Egypt, across the 1870s, relied on burning biofuels in the fireboxes of steam engines (Chapter 9) because imported coal was too expensive to allow for profitable steam-powered production cycles.

104 Henry Bernstein, 'Notes on Capital and Peasantry', *Review of African Political Economy* 10, 1977, 60–73.

8

The Chopping Axe Spring of Qau (1865)

In the early days of March 1865, peasants rose up between southern Assiout and northern Girga, just south of the great Rhoda sugar factory and extensive Daira Saniya plantations in Middle Egypt. It took Khedive Ismail three weeks, thousands of soldiers, and several steamers to put a decisive end to the uprising, by means of razing several villages to the ground, killing thousands of suspected rebels, and displacing thousands more. From this short-lived and fleeting but seemingly extensive uprising, the relevant records that survived are rare, and the events that transpired across time offer conflicting motives.

This chapter serves two purposes. The first purpose is to explain how the uneven process of transition to fossil-powered technology in agrarian production in association with the Delta's cotton boom in the early 1860s had an impact on Upper Egypt in the form of an extensive uprising that claimed thousands of victims. This means historicising the bigger socioecological, economic, and political processes associated with bringing this technology on an unprecedented scale (Chapters 5 to 7), as well as those resulting from the sudden and expansive deployment of this production technology, including in the wider circle in which it was embedded. For example, in the 1860s, Middle and Upper Egypt – before sugar production there was revolutionised by the early 1870s – came under more strain to produce food for the Delta, besides Cairo and other metropolitan spaces, when it turned overnight to cotton, which replaced food crops, as steam-powered cotton gins and irrigation pumps propelled the expansion in growing cotton on more Delta land. Just one impact was the Ibrahimiya canal, which was dug between 1865 and 1873, a megaproject of over 240 kilometres between Assiout and Bani-Suef, intended to serve the massive Khedival landholdings in Middle Egypt, to grow food crops that cotton replaced in the Delta, then to grow more sugar cane on those lands and replicate the lucrative profits of cotton. Corvée recruitment from Girga for this project might have been (as we will see shortly) the immediate trigger for the Qau uprising. Before that, the death of animal power in Middle and Upper Egypt because of the cattle murrain – which was tied intimately to the cotton boom – was another probable trigger. The taxing of peasants from the region in camels and recruits for the Sudan campaigns Ismail dispatched to Sudan early in his career to grow cotton was another. The causal chains between the uprising and the transition to the steam engine are long and complicated, as the chapter reveals, and highlighting some of them upstream above Cairo contributes to revealing how the steam engine found its way to the Nile valley.

The second purpose is that the Qau uprising's story needs to be told in detail, as it is a nearly completely forgotten memory in modern Egyptian history. So far, it is rarely brought up or mentioned in modern Egypt's histories. This chapter will hopefully bring it to life and tie it to the context of the transition to steam-powered technology.

Martina Rieker provides the most nuanced framework we have for understanding this 1865 Upper Egypt uprising.[1] She proposes three broad categories of motives reported from the period, which intersected to varying degrees with the land dispossession and proletarianisation questions. The first, that which seems to have overriding support from historical records on landownership and labour patterns, is the 'labour dispute thesis',[2] where mounting pressures on landless peasants and small landowners – those with properties under three feddans – led to the uprising. The second version is the one from the British officials and British newspapers. There, labour unrest is explicit, but framed as

1 Martina Rieker, 'The Sa'id and the City: Subaltern Spaces in the Making of Modern Egypt', PhD diss., Temple University, 1997. Also see Zeinab Abul-Magd, *Imagined Empires: A History of Revolt in Egypt*, Berkeley: University of California Press, 2013, Chapter 4.

2 Rieker, 'The Sa'id', 85.

'maladministration', where corvée and Khedival land confiscations distorted proper capitalist business conduct, according to the British officials in Egypt and the British press. It supported the claim that the uprising was sustained by the proletariat – the corvée-susceptible populations – but also pushed for linking Ahmad Al-Tayeb, the figure that several sources converged on as the leader of the uprising, to the Indian Rebellion of 1858 against the British East India Company. It presented a narrative where this Mahdi (messianic leader) planned a massacre against the Christians of Upper Egypt, both local Copts and foreign 'Franks'.[3] The version of the story that contradicted the labour question the most was the third one, which came from the American Presbyterian missionaries in Assiout, but even that one included evidence on underlying labour and socio-economic grievances, in spite of focusing mainly on a 'violence against Christians' motive.[4]

Based on Rieker's categories and on primary records, let us restate the rubric of the story in simpler terms: 1) the purely labour dispute thesis, from the landholding records; 2) the maladministration of labour and land/Indian connection thesis, following the British official story; 3) the religious-conflict thesis, with underlying labour grievances, from the Presbyterian missionary records. The story that follows factors in the three categories. However, a layer of complexity absent from this is integrating the events of the cattle plague and the moment of transition to steam cultivation, in association with Ismail's predatory designs on Sudan and the means by which he exploited Middle and Upper Egypt's humans and animals to that end. This is the fourth horizontal aspect I add to Rieker's rubric, in the sense that these three kinds of triggers were operating on it rather like a stage: the structural context in which the people of Qau gathered the discontent to rise. Let us call it the cotton boom/steam cultivation aspect for short. We will see how accommodating the transition to steam power tilts the balance greatly towards the labour-unrest narrative.

3 Ibid., 111–15.
4 Ibid., 89–92.

The Murrain, the Discontent, and the Messiah

Lady Duff-Gordon's letters are central to this attempt at bringing the Qau uprising of 1865 to life, for several reasons. Her work was one of the least racist and colonial travelogues one can find from European and American travellers in Egypt around the third quarter of the nineteenth century. More than any others found, by far, her work brings forward the voices of the subaltern at the centre of the uprising. In 1861, her doctors recommended that she move from England to a dry climate to heal from tuberculosis, so she went first to South Africa, before moving in 1862 to Luxor in Upper Egypt, which she found more conducive to healing. She returned to England briefly for the summer of 1863, then late in the year, she again returned to Luxor, where she was to stay for the winter of 1864 while her family remained in England, to continue recovering from her respiratory condition. She was a medical refugee of sorts who made friends with the locals, learned the language, and was sympathetic to their ordeals. Through her correspondence, she followed the progress of the uprising meticulously, recording reactions, news, and rumours from her many friends in Luxor, including government officials, middle and big landholders, and other European tourists and friends. She witnessed events leading up to the uprising, and recorded information on the destinies of the vanquished rebels and peasants who happened to live at ground zero. As such, to examine her letters closely is to follow the most detailed account that survived from the Qau uprising. I use other sources too, including the British press, some relevant Khedival orders, oral narratives, information from European travellers, and the memoirs of Nubar Pasha. Egyptian archives potentially hold relevant records, but they remain 'impossible archives'.[5] The progression of the cattle plague south of Egypt and processes associated with the cotton boom, which we can glean from her letters in this section, are a case in point on why this strong reliance on her letters is justified.

On 17 December 1863, Duff-Gordon hitched a ride from Cairo to Luxor, on the steamer that towed the *dahabiya* of M. Mounier, the agent

5 Pascale Ghazaleh, in *Search of the Archive: On Writing History in Egypt Today*, Berlin: Forum transregionale Studien, 2017; Hanan Hammad, 'Daily Encounters That Make History: History from Below and Archival Collaboration', *International Journal of Middle East Studies* 53, 2021, 139–43.

of Prince Halim who oversaw his agrarian interests in Upper Egypt. Her letters to her husband and mother and a few others during the trip offer us a credible thermometer on the cattle plague, cholera, and discontent across Upper Egypt that winter and the next (1865).

Before 3 January 1864, M. Mounier's steamer, on which she took up a room, was going upstream from Minya to Assiout. There, she wrote in a letter to her husband that the cattle plague devastated Bani-Suef, rendering irrigation a huge challenge, but did not yet reach Minya.[6] By the end of the month, before 29 January 1864, the plague had swept through Minya, potentially the northern tips of Girga as well, but not Esna (including Luxor), and food prices were still normal.[7] By late February, the impact of the dire situation in the north was spilling over as far south as Luxor, even if the plague had not yet crossed the Assiout–Girga line:

> I fear the loss of cattle has suspended irrigation to a fearful extent, and that the harvests of Lower Egypt of all kinds will be sadly scanty. The disease has not spread above Minieh, or very slightly; but, of course, cattle will rise in price here also. Already food is getting dearer here [. . .] and bread [price] has risen considerably.[8]

The new corn harvest of late March 1864 around Luxor was 'unusually fine', as Duff-Gordon heard, but the murrain was already invading the lands north of the Minya/Assiout line, and 'the destruction has been fearful'.[9] In early April 1864, the cattle disease was still absent that far south, but 'a good deal of sickness among the people' was spreading.[10] The cholera came; she described it as a strange epidemic a week later, 'the people get stoppage of the bowels and die in eight days unless they are physicked', and the Lady started practising medicine. One reason was that the peasants did not trust the government physicians.[11]

The next week, on 23 April 1864, the cattle plague was in full swing in Luxor and south of Luxor, at Muta'na, where the Mouniers, the Lady's

6 Lucie Duff-Gordon, *Lady Duff Gordon's Letters from Egypt*, New York and London: McClure, Philips & Co., R. Brimley Johnson (revised edition), 1902, 91.

7 Ibid., 108–9.

8 Ibid., 122.

9 Ibid., 140.

10 Ibid., 156.

11 Ibid., 157.

friends, were staying on their large estate, where she heard that 'men and cattle' suffered death and disease. Muta'na, with a human population of 2,000 at the time, lost 150 heads of cattle, and eight to ten human deaths were reported per day.[12] By early May 1864, the situation became worse. M. Mounier buried over a thousand heads of cattle, and forty of his men were dead from the cholera, while 'dead cattle float down [the Nile] by thousands'.[13] When she returned to Luxor by mid-May, she saw that the outbreak was complete:

> All the cattle were dead. Mustapha lost thirty-four, and has three left; and poor farmer Omar lost all – forty head. The distress in Upper Egypt will now be fearful. Within six weeks all our cattle are dead. They are threshing the corn with donkeys, and men are turning the *sakiahs* – water-wheels – and drawing the ploughs, and dying by scores of overwork and want of food in many places. The whole agriculture depended on the oxen, and they are all dead. At El-Moutaneh and the nine villages round Halim Pasha's estate 24,000 head have died; four beasts were left when we were there three days ago.[14]

On 2 June 1864, around fifteen months after the murrain showed up in the Delta, nearly all the cattle in Luxor were dead, Duff-Gordon reported – no more than five were still alive. Her friend Omar had one left from among fifty-four heads. Local grain stayed after harvest in unprocessed heaps on the fields, and famine food prices became the standard. No mail was coming from Cairo as the Nile was low, 'and now I really know how Egyptians live'.[15]

And under these famine food price conditions, around mid-August 1864, Greek moneylenders saw the opportunity. They came tailing the tax-gatherers,

> like a vulture after a crow [. . .] now wheat is at 170 piasters the *ardeb* here, and the fellah has paid 3.5 per cent a month besides [in loan

12 Ibid., 163.
13 Ibid., 166.
14 Ibid., 169.
15 Ibid., 179.

> interest to the Greek creditors]. Reckon the profit! Two men I know are quite ruined, and have sold all they had. The cattle disease forced them to borrow at these ruinous rates, and now alas, the Nile is sadly lingering in its rise, and people are very anxious.[16]

Borrowing became fractal. On the scale of the state, Ismail borrowed European financial capital to address the epizootic, after it visited Egypt because of his efforts towards working the lands he acquired once he ascended the throne. On a smaller scale, farmers, likely middle and small farmers, borrowed from petty usurers under the impact of the epizootic crisis that visited their animals.

Between late August 1864 and January 1865, Lady Duff-Gordon was absent from Luxor. This was also around the time when trouble was brewing in Sudan after two years of Khedival 'improvement' projects there, that was likely inspired by the cotton boom. The second day Ismail started sending out orders, in January 1863, right after he ordered his first 1,000 oxen from Sudan, he sent another order to his men there asking them to 'improve the status of the country and its people'. To increase their wealth and means of sustaining livelihood, mainly through 'cultivating cotton [. . .] which is among the reasons for increasing wealth and welfare'.[17] In other words, give the people of Sudan incentives, across all provinces under Viceroyal control, to cultivate this plant.[18] These efforts to improve the status of Sudan and its people led to a popular uprising in the White Nile province, then a mutiny among the troops, or a series of mutinies, around the time Ismail sent another order for the soldiers to grow the cotton themselves, while the Sudanese were getting educated and civilised.[19] This last order came in January 1865, accompanied by a detailed bylaw for organising cash-cropping and educating the people of the White Nile province in Sudan. Perhaps the troops that were sent to the White Nile were those who mounted camels confiscated from Lady Duff-Gordon's neighbours in Luxor a few days before the bylaw on cultivating cotton in the While Nile was issued in late January.

---

16 Ibid., 190–1.

17 Amin Sami, *Taqwim Al-Nil 3.2* [Survey of the Nile Vol. 3.2], Cairo: Matba'at Dar Al Kutub, 1936, 448–9. Emphasis added.

18 Ibid.

19 Ibid., 563–7.

January 1865 started for the Lady with a dismal and extraordinary sight. In front of her house in Luxor, 'the sand [. . .] is thronged with all the poor people with their camels, of which the government have made a new levy of eight camels to every thousand feddans'.[20] The animals were taken to transport soldiers to Sudan, where they – the camels – would have eventually died, since they were not accustomed to the desert, she heard. And she found the camel-confiscation process shocking:

> The whole place is in desolation, the men are being beaten, one because his camel is not good enough, another because its saddle is old and shabby, and the rest because they have not money enough to pay two months' food and the wages of one man, to every four camels, to be paid for the use of the Government beforehand. The *courbash* [whip] has been going on my neighbours' backs and feet all the morning. It is a new sensation too when a friend turns up his sleeve and shows the marks of the wooden handcuffs and the gall of the chain on his throat. The system of wholesale extortion and spoliation has reached a point beyond which it would be difficult to go.[21]

She speculated about why the troops who rode these camels and others taken from across Upper Egypt were going to Sudan:

> I fancy from the number of troops going up to Assouan that there is another rising among the blacks. Some of the black regiments revolted up in the Soudan last summer, [. . .] and the camels are being sent off by hundreds from all the villages every day.[22]

Those same troops that were travelling to Aswan then onwards to Sudan in early 1865 might have had among their ranks some of the soldiers who rebelled a few months later. For, on 23 October 1865, Ismail ordered his commander in Sudan, Ga'far Pasha, to take decisive action against the hundreds who had recently mutinied, and he sent detailed

---

20 Duff-Gordon, *Lady*, 208.
21 Ibid.
22 Ibid., 209.

instructions on punishments. Since the crime was grave, 'there is no room for investigation' – 118 soldiers, including some officers and ranking officers, 'should be instantly executed in any way' and their leader kept 'on a cross for 24 hours to make an example of him'. For the remaining 635 soldiers and officers who were arrested, 'one in each nine shall be executed, while the remaining eight shall be beaten on their feet by chains and used in hard labour without pay. Only provide them with food.' The guards who were to supervise the prisoners 'shall have their guns loaded, and kill instantly any of the mutinied [i.e. those punished with hard labour] who express the slightest dissent, physical violence, or verbal aggression'.[23]

Moving back in time to Luxor in early January 1865, Lady Duff-Gordon summed up the suffering of the peasants taxed in camels in Upper Egypt, in the same letter to her mother:

> From my window now I see the men limping about among the poor camels that are waiting for the Pasha's boats to take them, and the great heaps of maize which they are forced to bring for their food. I can tell you the tears such a sight brings to one's eyes are hot and bitter. These are no sentimental grievances; hunger, and pain, and labour without hope and without reward, and the constant bitterness of impotent resentment. To you all this must sound remote and almost fabulous. But try to imagine Farmer Smith's team driven off by the police and himself beaten till he delivered his hay, his oats and his farm-servant for the use of the Lord Lieutenant, and his two sons dragged in chains to work at railway embankments – and you will have some idea of my state of mind today.[24]

And let us not forget that Farmer Smith of the Nile valley was not only suffering from taxation in crops, farm animals, or money, or the indentured labour of his sons. It was the same time that Ismail was confiscating lands at will across the country. Smith Effendis were turning into landless labourers.

How did the people react? 'Last week [2 January 1865] the people were cursing the Pasha in the streets of Assouan, and everyone talks

23 For all citations in paragraph, Sami, *Taqwim Vol 3.2*, 625.

24 Duff-Gordon, *Lady*, 209.

aloud of what they think', and in the field, they sang songs of hunger and great distress:[25]

> The song of the men watering with the *shadoofs* sounds sadly true as they chant *Ana ga-an*, etc. 'I am hungry, I am hungry for a piece of dourrah bread,' sings one, and the other chimes in, *Meskeen, meskeen* 'Poor man, poor man,' or else they sing a song about Seyyidna Iyoob 'Our master Job' and his patience. It is sadly appropriate now and rings on all sides as the *shadoofs* are greatly multiplied for lack of oxen to turn the *sakiahs*. All is terribly dear, and many are sick from sheer weakness owing to poor food.[26]

But it did not stop at camels confiscated from their owners, or men replacing the oxen on the irrigation works, or the proletarianisation of small landholders immediately before the plague, or potentially recruiting even more peasants as soldiers to make up for those who were sent to Sudan in January to cultivate cotton. In February 1865, Ismail taxed Upper Egypt 50,000 men in corvée, outside the usual quota, which included those taken to dig the Suez Canal. Those 50,000 men were taken to dig the Ibrahimiya canal, the Lady heard.[27] When it was completed by 1873, the canal extended for over 200 kilometres, with the width of 60 metres along one third of this distance, and 15 metres for the other two thirds.[28]

The land was buzzing with anger. The poorest of the peasants, the most hungry, the men working the *shadufs* and tied to the waterwheels, were certainly those who suffered the most, but the bigger landholders suffered also through the loss of camels and of their labourers, who died of hunger, weariness, and sickness from the cholera, or were recruited for the construction of the new irrigation canal. We might never know, but perhaps this great corvée recruitment in February 1865 for Ibrahimiya was the last straw. The thermometer hit critical temperatures.

Suddenly, in March 1865, the crackling coals erupted in fire. Ten villages were now in open revolt, at the Assiout–Girga line, and Effandina

25 Ibid., 208.
26 Ibid., 211–12.
27 Ibid., 212; 'Occasional Notes', *Pall Mall Gazette*, 29 March 1865.
28 J. C. McCoan, *Egypt As It Is*, New York: Henry Holt and Company, 1877, 235.

the Viceroy had come to the south himself. The people rose, and the revolution of Qau came, heralded by the communist messiah of the fellah, Ahmad Al-Tayeb:

> Sheykh Yussuf returned from a visit to Salamieh last night. He tells me the darweesh Achmet et-Tayib is not dead, he believes that he is a mad fanatic and a communist. He wants to divide all property equally and to kill all the Ulema [clergymen] and destroy all theological teaching by learned men and to preach a sort of revelation or interpretation of the Koran of his own. 'He would break up your pretty clock,' said Yussuf, 'and give every man a broken wheel out of it, and so with all things.'[29]

Before we proceed to the messiah of Qau, who was about to break clocks and redistribute all things equally, let us recap this avalanche of events picked from Duff-Gordon's rich and detailed letters of 1864 and 1865, and what we knew about the cattle plague and flood in the preceding chapters. We will stop at why the Qau uprising erupted, before we proceed to how it unfolded.

Two months into his reign, in March 1863, Ismail Pasha tried to quarantine the cattle plague that was witnessed first, in March 1863, in Behira, north-western Nile Delta. Simultaneously he ordered more oxen, including from Turkey and beyond, as he was planning to expand – on his own proliferating lands – in growing cotton, given the cotton famine brought about by the American Civil War. The Turkish oxen he imported were infected by the cattle plague and spread the deadly virus across all animal power, including upstream beyond the cotton plantations of the Delta to engulf the Nile valley by 1864. The cotton enterprise indeed collapsed for a short time in the Delta as it fell victim to the plague and the high flood of 1863, so the Pasha turned his eye to the south: Middle and Upper Egypt, and Sudan. He taxed the people of the former in camels to take the troops to the latter. The troops were to oversee and take part in the cultivation of cotton in Sudan, but both the people of Sudan and the soldiers sent to develop them into cash-croppers rebelled. Whether the Sudan cotton plans were inspired by the epizootic is not important to discern, since the logic of expansion in

29 Duff-Gordon, *Lady*, 217.

cotton – which brought the epizootic to the country – was ultimately propelled by the urge to make profits out of the cotton famine. Meanwhile, as the murrain and cholera were collecting their heavy toll, he taxed the poor peasants of Middle and Upper Egypt in corvée to build an irrigation canal that was to serve as infrastructure for the expanding – at the time – Khedival lands, the Daira Saniya. This patch of land had grown by the 1870s to around 182,000 acres administrated centrally by the *Daira* administration that formed a mini state apparatus just for this chunk of land south of Cairo, and was largely centred on the production of refined white sugar.[30]

Behind immediate triggers of the uprising stood the structural grid of processes of which these events were only the most recent manifestations. The list is long, and it sums up Ismail's measures for cash-cropping over his first two years in office. It includes – not a conclusive inventory – intensified corvée labour, land dispossession on a grand scale, and the progress of monetisation, accompanied by penetration of European entrepreneurs and usurers deep into Upper Egypt.

March 1865: The Qau Uprising

The uprising takes its name here from its geography. Martina Rieker reconstructed the boundaries of the insurgent zone from *Al-Khitat Al-Tawfiqiya*, the twenty-volume geographical encyclopaedia on Egypt by Ali Mubarak (first edition 1889). She traced mentions in the encyclopaedia's entries on villages and hamlets to roles in the uprising. According to this mapping, the village at the centre of the storm, from where the peasants rebelled, was Qau Al-Gharb (Qau to the west of the Nile). It was also the only village involved to the west of the river. East of the Nile, across from Qau Al-Gharb, was the 'Uqal basin. This fertile area is wedged between the Nile to its west and the villages on the Qau irrigation canal to its east. East of the Qau canal is the eastern desert. The 'Uqal basin villages were reported to have joined Qau Al-Gharb wholeheartedly: Riyania, Natra, and Sheikh Jabr. Slightly to their east were many villages, big and small, extending south to north along the Qau canal, which joined in to varying degrees. From south to north, east of

30 McCoan, *Egypt*, 147.

the basin, the smaller hamlets of Kafr Al-'Uqal and Nazlet Al-'Alam joined, as well as the bigger villages of Al-'Uqal Al-Qibli (southern 'Uqal) and Al-'Uqal Al-Bahari (northern 'Uqal).[31]

A relevant question here is, given that these structural processes and events that unfolded right before the uprising were generally plaguing Middle and Upper Egypt that year, why did the revolt come out of the Assiout/Girga borderline in particular? Landholding and proletarianisation patterns across this geography offer leads on why this zone held more subversive potential. Rieker's detailed analysis of landholding patterns across the uprising's zones, identified previously, pushes towards the conclusion that the people of this zone could have been susceptible to corvée more than those upstream and downstream, and this corvée vulnerability within the area must have varied depending on where more landless and small landowners were living in greater concentrations.[32] Large Viceroyal plantations and powerful big landholders – to the north and south of our zone – needed hands to work their land full-time at all seasons, and they must have wanted them to stay there permanently and had power that matched the desire to see this through. But the landless labourers and reservoir of small landholders at the Assiout/Girga line were where one would have found many concentrations of those who can be spared without compromising cash-cropping on the great plantations. Who would have been taxed in labour (corvée) and sent to far-off places, in bigger proportions than others, if not the landless masses of this pocket between large estates to the north and south? This geography of dispossession must have also been why Ismail reacted severely and came himself at the helm of a big army to terminate the uprising before it spilled over to the north or south, where his interests would have been directly compromised had the fire caught his *Daira*. A pocket of potentially recurrent resistance, threateningly pulsating in the middle of lucrative production sites, was to be made an example of. It needed to be swept clean and razed to the ground.

According to several sources, the first event of the uprising, and the one that awakened Cairo and pushed her to send brute force, was an attack on a steamer going downstream, north of our site. A letter came

31 Rieker, 'The Sa'id', 74.

32 Ibid., 74–84.

to the *Pall Mall Gazette* in London in late March 1865, 'from a well-informed native gentleman at Cairo', saying that Sheikh Ahmed Al-Tayeb, after he collected around him 1,500 followers who took him as their spiritual leader, planned the takeover of a steamer. He accompanied his men to a spot on the eastern bank of the river, across from Aboo-Hannes, a site described as somewhere north of Qau, between Rhoda and Radamon, sites of extensive sugarworks in northern Assiout since the 1830s. It was a spot where a boat after losing steering would be swept, under strong current, to the eastern bank, where the rebels waited.

> A steamer soon appeared making its way down stream, when suddenly several shots were fired into her, and the helmsman fell. The vessel swung round with the stream, and was carried right into the hands of the insurgents, who pillaged her forthwith, getting booty, in money alone, of about [£]3,000.[33]

The incident of the pillaged steamer first reached Lady Duff-Gordon much exaggerated. In the beginning, she heard it was a Prussian boat, and 'all on board murdered, and the boat burned'.[34] Then in the same letter to her mother: 'it seems that the boat robbed belonged to Greek traders [who came with intensified monetisation in Upper Egypt along with cash-cropping and/or to extend credit after the loss of cattle to the murrain], but no one was hurt, I believe, and no European boat has been molested'.[35] Two weeks later, she wrote to her husband that only the steersman was killed.[36]

If it was, indeed, an uprising that started a few kilometres to the south, at Qau Al-Gharb and the 'Uqal basin, rather than a singular highway robbery event, then what happened before the steamer event to the north of our site? And if Al-Tayeb was there among many followers, how did he rally them around him in the first place? A traveller in Upper Egypt through March 1865, who reportedly returned late in the month to Cairo,

33 'Troubles in Egypt', *Pall Mall Gazette*, 1 April 1865.
34 Duff-Gordon, *Lady*, 215.
35 Ibid., 216.
36 Ibid., 220.

> report[ed] a serious disturbance in that country. The villages twenty miles north and south of Kao [. . .] are said to have revolted against the Government on account of having been forcibly taken to work on the Viceroy's estates, at nominal wages, during the crop season, *and then compelled to dig at a large canal which his Highness is making in that district* [the order requesting 50,000 corvée labourers in February 1865] of course, equally without wages. It is also said that a man of the neighbourhood has *lately proclaimed himself as the herald of the Messiah*, who is to come from the East, and who is to be the leader of Moslems as well as Christians [. . .] Whatever the precise reasons, a number of villages turned out and opposed the Government officer.[37]

Evidence from Lady Duff-Gordon, alongside that from Ali Mubarak in Rieker's account, does not indicate that the uprising extended for forty miles around Qau. Maybe it spilled over, according to this traveller cited by the *Pall Mall*? Perhaps the rebels of Qau were marching up and down, rallying the people of more villages to the uprising? The Lady apparently asked her friends in Luxor how one becomes a messiah (a Mahdi), and she got an answer on the tradition according to the Sufi *shareef* Luxorians. Ahmed was, in fact, not the first Mahdi in the family:

> I will tell it as I heard it. He did as his father likewise did thirty years ago, made himself *Ism* – name – by repeating one of the appellations of God, like *Ya Latif* three thousand times every night for three years which rendered him invulnerable. He then made friends with a Jinn who taught him many more tricks – among others, that practised in England by the Davenports of slipping out of any bonds. He then deluded the people of the desert by giving himself out as *Al-Mahdi* – he who is to come with the Lord Jesus and to slay Antichrist at the end of the world – and proclaimed a revolt against the Turks [i.e. central authority].[38]

So far, the story appears like something out of *One Thousand and One Nights*, but messianic uprisings in nineteenth-century Egypt were

37 'Occasional Notes', *Pall Mall Gazette*, 29 March 1865. Emphases added.

38 Duff-Gordon, *Lady*, 215.

standard practice among dissenting peasants before angry masses articulated their uprisings in the language of the state and modern insurgent politics two decades later.[39] We have on record from Rieker at least four earlier messianic uprisings between the turn of the nineteenth century and its first quarter: Muhammad Al-Kitani reportedly led one in Qina and Girga in 1798. The next three were against Mehmed Ali, whose reign was pierced by recurrent local uprisings across the country, especially across the 1820s. In Upper Egypt alone, in 1820–1 Ahmad Al-Salah was the Mahdi in Qina. Then came Ahmed Al-Mahdi Al-Wazir in 1822–23 in an apparent extension of the 1820–1 revolt, which spilled over from Qina to Girga to its north. Then it was apparently carried forward in 1824 by Ahmed ibn Idris across the same districts.[40] Other local uprisings were termed as 'spontaneous' or leaderless, between 1777 (at Tahta, southern Girga) and 1838–9 (same zone), and were generally caused by hunger or harsh conscription practices.[41]

Khaled Esmail, a novelist from Koum Al-Arab, a village neighbouring Qau and the 'Uqal basin, offers a complementary perspective on Ahmed Al-Tayeb, the messiah of 1865, and his predecessors. He charted a pattern of resistance grounded in the socio-economic and political structures of the nineteenth century, not the kingdom of heaven. According to Esmail, there was a long line of dissent from the *ashraf* – plural for *shareef*; descendants of Prophet Mohamed – of Upper Egypt that started in 1809. Our Ahmed Al-Tayeb of 1865 was indeed a *shareef*. During the period of great disturbances after the French invasion ended disastrously for Napoleon's army (1798–1801), the *Naqib* (supreme leader) of the *ashraf*, Omar Makram, who was living in Cairo at the turn of the century, led the civil disobedience and unrest that brought Mehmed Ali to power in 1805. Then, in 1809, when Makram advised

39 This articulation around freedoms, representation (*Shoura*), and the nation came strongly during the Urabi Revolution of 1881–2. This language was used profusely in speeches and newspaper articles by the second man in the revolution, and its mouthpiece, Abdullah al-Nadeem. A generous sample of such articles and speeches can be found in Ali Al-Hadidi, *Abdallah Al-Nadeem: Khateeb Al-Wataniya* [Abdullah Al-Nadeem: The Orator of Patriotism], Cairo: Al-Mu'athatha Al-Misriya Al-'Amma, 1962.

40 Rieker, 'The Sa'id', 69.

41 Ibid. More on local uprisings in the early nineteenth century, and briefly on Qau: Judith Tucker, *Women in Nineteenth-Century Egypt*, Cambridge: Cambridge University Press, 1985, 139–43.

the Pasha to reduce the taxes recently imposed on the people lest a revolution erupt, the latter banished him to Damietta at the northern tip of the Delta. Resentment grew between the *ashraf* and Mehmed Ali. At the time, the Pasha's taxation and land-redistribution measures also disadvantaged them greatly, as he gave privileges in land and status to rivalling Bedouin groups to settle and administrate taxation in the south, so some of the *ashraf* became magnates to which popular discontent was attracted in times of trouble.[42]

Let us look at an example, which perhaps is another narrative of the series of early-1820s messianic uprisings listed by Rieker previously. In 1820, Ahmed Ebaid Al-Ga'fari of Aswan, a Sufi leader and a *shareef*, moved to Qina. There, he took a leading role in the early-1820s insurgencies of Qina and Girga but had to flee the Pasha's wrath to Iraq eventually.[43] About this messianic 1820s insurgency, Khaled Esmail cited Ali Mubarak's *Al-Khitat*:

> A man came forth from Salimiya [Ahmed Al-Tayeb's hometown], Sheikh Ahmed [Ebaid Al-Ga'fari]. People rallied around him, and he took their pledges [of loyalty]. His followers reached 40,000, and as such he took pride in numbers and rebelled against the government. From among his followers, he appointed rulers of districts. His criminality spread across the land. He pillaged governmental monies and granaries, and confiscated the money from the tax collectors. He corrupted the land and the Nile, and the country and its rulers feared him.[44]

When he returned from Baghdad years later, he took up the name of Al-Bughdadi and established a village by the name Al-Bughdadiya, before resuming his role as a spiritual leader of great influence in his immediate geography.

Khaled Esmail confirms further that, although the oral tradition in Upper Egypt is consistent with the story on Al-Ga'fari-then-Bughdadi's role in the unrest of the 1820s, the official narrative from Ali Mubarak

42 Khaled Esmail, 'Thuwar wa Manfiyoun: Min Ayn Yastamid Al-Sheikh Al-Tayeb Quwataho? [Rebels and Exiles: From Where Does Sheikh Al-Tayeb Draw His Power?]', *Al-Manassa*, 3 December 2020.

43 Ibid.

44 Cited in Esmail, 'Thuwar'.

omitted the fact that such insurgencies usually spilled over to Girga and Assiout in the north.[45] So there seems to have been already a long legacy of trouble between the government in Cairo and some of the *shareef* Sufi leaders of Upper Egypt fuelled by socio-economic grievances. The natural leaders of Qina – rich landholding *shareef* families centred around Luxor, Salimiya included – were there whenever popular unrest erupted. Perhaps Ahmed Al-Tayeb was just the next leader in this long line of messiahs, redeemers of the people and subduers of *al-masih al-dajjal* (Antichrist), that is, the central government sucking the country dry, to the last ox, camel, or human still standing.

Qau Al-Gharb, largely composed of small landholders and the landless, must have easily answered the call to defy the 50,000-corvée order hanging heavily over their heads. The three villages of the 'Uqal basin's villages were no less responsive. These were, indeed, the villages Lady Duff-Gordon heard were revolting, 'three villages below [north of] Keneh – Gau, Rayanaeh and Bedeh [another name for the Sheikh Jabr and/or Natra]'.[46]

We previously left the rebels at the pillaging of the Greek steamer. *Reuter's Express* landed on robbery as the main motive of the uprising, saying that local senior officials of Assiout and their law enforcement troops instantly descended on the rebels after they telegraphed Cairo, but it did not bring up the steamer incident.[47] Then, under command of the Assiout governor, in a first confrontation, 'the robbers were surrounded, and in a skirmish about 10 of them were brought to the ground'.[48]

According to the same news story, Cairo responded and quickly dispatched a certain Schiahim Pasha – alternatively, in other sources given here, Sherin Pasha, but his actual name was Shahin Pasha[49] – who came

45 Ibid.

46 Duff-Gordon, *Lady*, 215.

47 'Egypt', *The Morning Post*, citing: *Reuter's Express*, 6 April 1865.

48 Ibid.

49 In Nubar Pasha's memoirs, a certain Shahin Pasha was the leader of the first batch of troops sent south to police the uprising. See Nubar Pasha, *Muzakirat Nubar Pasha* [Arabic translation by Jarou Robert Tabaqian of: Mémoires de Nubar Pacha], Cairo: Al-Shorouq, [1983] 2009, 324. Also spelled 'Shahin Pasha' in Andrew Watson, *The American Mission in Egypt: 1854–1896*, Pittsburgh: United Presbyterian Board of Publication, 1904, 179.

> some days afterwards [after the first skirmishes]. He had been despatched in all haste, landed with troops of the regular army. The brigands, surrounded by fire and sword in a contracting circle, were very soon compelled to lay down their arms. About 20 of the depredators who were known to be the chief instigators of these disturbances were arrested, and, after trial, will be banished to a province of Upper Soudan. The prompt and energetic measures resorted to had the effect of restoring perfect tranquillity in the town of Siout [Assiout] and the country in the neighbourhood.[50]

The same account tells us that Shahin Pasha was wounded, and that the *mudir* of Gerza was assassinated by the rebels. In addition, '400 camels laden with sugar belonging to his highness the Viceroy had been plundered'.[51]

The *Pall Mall* heard another version of the story on the first skirmishes that presumably ended in a hard defeat for the followers of Al-Tayeb. Sherin Pasha (Shahin) came on board a steamer from the north for a quick showdown with the rebels, while they were still busy with the spoils of the attacked Greek steamer,

> and, quickly seeing the state of affairs, opened fire on the Sheikh and his accomplices. These latter lost no time in beating a retreat, and took up a clever position on the brow of a hill, so as to be beyond the reach of a charge of cavalry which had opportunely arrived. Sherin Pasha, who had come up in the steamer, had with him two small guns, and as soon as he could place his troops in position ashore, ordered one of the guns to be aimed at the standard, about the middle of the Sheikh's mass. Down came the standard, and a grape shot is said to have pierced the Sheikh's back, which compelled him to be carried higher up the hill. When laid in a cave, out of sight of the Government troops, he announced to his attendants that he would shortly be called to another world in order the better to inspire and guide his disciples, and then died. This intimation, however, failed to inspirit those present, who dispersed in every direction. Some were cut down, some

50 'Egypt', *The Morning Post*, 6 April 1865.
51 Ibid.

> shot or drowned, and other hung or beheaded. The casualties on both sides amount to about sixty in all.[52]

But if it was Shahin Pasha's armed force that killed Al-Tayeb and his followers, why did Fadhl Pasha follow and why did Ismail travel there, as we will see shortly? If this latter piece of news is true, then the insurgency carried on until these two great men came from Cairo with much greater force a few days later. Maybe the Messiah's death inspired more insurgency? Perhaps he lived to fight the next day? Or, maybe, Ismail came nonetheless to make an example of the rebelling villages. The *Friends of India and the Statesman* actually cast doubt on the story of the early defeat: 'the Governor of Girgeh failed to put down the revolt at first, but troops were sent up the Nile in five steamers, the Pasha [Ismail] himself went, and eight hundred rebels perished'.[53]

He died, he did not die, alternately went the rumours. A British tourist passing through the district at the time – Rev. Smith, more about him in the next section – was told that Al-Tayeb's body was hung on a tree for days for people to see, to despair in the prospects of the uprising, but they simply said he could not die. Others said his body was thrown into a well, but the water rose up and delivered him to the surface and he escaped.[54] The same transpired at Luxor, according to Duff-Gordon. First, she heard on 13 March that 'some say [Al-Tayeb] is killed, others that he is gone off into the desert with a body of bedaween and a few of the fellaheen from the three ravaged villages'.[55] Later she heard that he became invincible: 'where the bullets hit him he shows little marks like burns'.[56] Then the Lady's friend, Youssef, told her a couple of days later that the *darweesh* lived, and he spoke of the mad fanatical communist plot for wealth redistribution, breaking clocks, and splitting everything evenly among the people.[57]

Al-Tayeb became a legend, dead or alive. Even if he died, the radical redistribution of wealth he and his disciples preached, and which

52 'Troubles in Egypt', *Pall Mall Gazette*, 1 April 1865.

53 No title, *Friends of India and the Statesman*, 27 April 1865.

54 Alfred C. Smith, *The Attractions of the Nile and its Banks, Vol. II*, London: John Murray, 1868, 145.

55 Duff-Gordon, *Lady*, 215–16.

56 Ibid., 225.

57 Ibid., 217.

reached the ears of Youssef at Luxor, must have been very disturbing to Ismail Pasha and the nobility. Rumours came out that Al-Tayeb had learned to mobilise the masses from a sheikh who fled the failed rebellion of India some years earlier. The colonial *Friends of India* cited previously naturally saw the resemblance to the late rebellion against the East India Company: 'near Girgeh, lives a dervish whose father excited a revolt 30 years ago. In the usual style, this fanatic declared himself to be the Imam Mehdee, the hope of whom induced so many Bengaleen to join the late Sittana rebels.'[58] The *Pall Mall* chimed in, conjuring a connection, not a resemblance. Al-Tayeb had 'taken lessons from an Indian co-religionist who visited Kao before the mutinies [and as such] was seized with the idea that he was destined to bring mankind over to the one true religion, which was to be a purified Islam'.[59] But then Ismail heard about the India connection, and flipped it. When Lady Duff-Gordon learned his version coming from Cairo, it was the other way around:

> The Pasha accuses some Sheykh of the Arabs of having gone from Upper Egypt to India to stir up the Mutiny against us [the British]! *Pourquoi pas* to conspire in Paris or London? It is too childish to talk of a poor Saeedee Arab [a man from Upper Egypt] going to a country of whose language and whereabouts he is totally ignorant, in order to conspire against people [i.e. the British] who never hurt him.[60]

She laughed with Youssef at the idea. But Ismail must have cooked this up not just for 'European taste'. He must have known very well about the rebellious tradition of the Sufi Mahdis and the trouble they had made for his grandfather, Mehmed Ali, a few decades earlier, at moments when they became one with the angry peasants. Perhaps he hung on to the story of the India connection to cover the stench of the long trail of the dead his army left behind as he razed their villages after that steamer event, the failure of Assiout and Girga officials to squash the uprising, and the pillaged 400 Khedival camel-loads of sugar.

58 No title, *Friends of India and Statesman*, 27 April 1865.

59 'Troubles in Egypt', *Pall Mall Gazette*, 1 April 1865.

60 Duff-Gordon, *Lady*, 236. And this was 1865, seventeen years before the British occupation, which meant that no grand-scale explicit British harm had yet occurred.

The same might have been behind blowing another story out of proportion, on Muslim-Coptic hatred as the trigger of the revolt. This version of the events occurred in almost all British newspapers that reported on the uprising, but the missionaries of the American Presbyterian Church in Assiout must have been a central source even if not actively reporting it as such to the press. The *Sun* reported in mid-April that 'it is a well-known fact that over the whole Mahomedan [*sic*] world just now there are floating rumours and predictions of a coming ascendancy, during which the Crescent is to stand above the Cross, and the faith of the East is to revive again'. The people are 'discontented with the Government, and not well-disposed towards the Franks, for whose sake they know the cotton is grown'. Then when the news arrived of the first Cairo force that confronted the rebels under commander Shahin, it was said to have come 'just in time to save the bulk of the Christian population of the place'.[61] So it seemed that the rebels were supposedly targeting Christians of all sorts, foreign and local. The *Friends of India* stated it bluntly: 'he [Al-Tayeb] said that in March the Mussulman ought to slay the Christians, and accordingly the fellahs rose against the Copts'.[62] The *Pall Mall Gazette*, meanwhile, reported that Al-Tayeb beheaded two Copts who declined to join his followers.[63]

The head of the Presbyterian Mission in Egypt, from Assiout, Mr Andrew Watson, provided a rich account on this religious conflict. He started his story on the uprising by saying that it broke out because of the murrain, the 'terrible oppression', and the recent order of 50,000 corvée labourers. He twisted it, however: 'on such occasions [i.e. peasant uprisings] the venom and hate of Muslims are always directed against the so-called Christians [i.e. Copts, not true Christians to him]'.[64] According to Mr Watson, Al-Tayeb saw a Copt smoking during Ramadan, while Muslims were fasting, so he cursed him and unleashed his men who pillaged his home. A few days later, 'he and his followers laid hold of some Coptic girls and carried them off, a very common habit with Muhammadan saints'.[65] Then Lady Duff-Gordon heard about a Muslim girl and a Coptic man instead, but dismissed the story as the

61 'The State of Egypt', *Sun*, 15 April 1865.
62 No title, *Friends of India and the Statesman*, 27 April 1865.
63 'Troubles in Egypt', *Pall Mall Gazette*, 1 April 1865.
64 Watson, *The American Mission*, 178.
65 Ibid.

'religious legend connected with [this] business', adding elsewhere that 'there is no religion in the business at all [. . .] It was like our Swing, or bread riots, nothing more.'[66] What she heard on the Coptic male/Muslim female was the following:

> A certain Copt had a Muslim slave-girl who could read the Koran and who served him. He wanted her to be his Hareem and she refused and went to Achmet et-Tayib who offered money for her to her master. He refused it and insisted on his rights, backed by the Government, and thereupon Achmet proclaimed a revolt and the people, tired of taxes and oppressions, said 'we will go with thee.'[67]

It is Mr Watson from the Presbyterian Mission who gives us a lead on where this religious legend overlaps with the communist sheikh who was planning to redistribute all wealth. Before the insurgents' battle against Shahin Pasha's troops, he heard that they planned

> a sudden descent on Asyut, and arranged with their friends to aid the prisoners in Asyut jail to effect an immediate escape, cut the telegraph wire, and take possession of the town awaiting their arrival, when *they would kill all the Turks, Copts, and Franks, and divide their personal effects among themselves.*[68]

Regardless of whether the insurgents had planned a descent on Assiout or not, the people there seem to have heard about their agenda of wealth redistribution and started thinking about potential targets. What was the common denominator of Turks, Copts, and Franks? It was not religion but class. Almost all Turks (state officials and big landholders) and Franks (foreign traders) were among the richest. Meanwhile, some Copts could have expected to have been targeted: those who were wealthy. For example, we know from *Al-Khitat* that Qau Al-Kubra across the Nile from Qau Al-Gharb from where the uprising erupted was 'a wealthy, predominantly Coptic agricultural community'.[69]

66 Duff-Gordon, *Lady*, 226. Ibid., 222–3.

67 Ibid., 225–6.

68 Watson, *The American Mission*, 179. Emphasis added.

69 Cited in Rieker, 'The Sa'id', 78.

More importantly, the bigger landholders in northern Assiout and Minya farther north had among them many Coptic Pashas and Beys.[70] Other targeted Copts would have been government officials. We know that, since at least the seventeenth century, the civil servants doing the administrative part of the tax-collecting process, the *sarrafs*, were generally learned Copts, to the extent that a book from the early modern period uses *sarraf* and *nusrani* – another word for Copt – as synonyms.[71] Meanwhile, Lady Duff-Gordon, a Frank herself, refused an offer for protection by police guards as the uprising raged immediately to the north.[72]

For Lady Duff-Gordon, the Muslim/Coptic affair was a ridiculous game of 'divide and conquer' played by the central government, and she mocked the senior officer for it. Meanwhile, a Coptic friend told her that his uncle's family were all shot to death in Qau – though this could have been Qau Al-Kubra, the rich overwhelmingly Coptic community at the time, across from Qau Al-Gharb – when Ismail's punishing hand eventually reached it, and that all Copts and Christians there faced the same fate.[73] Were they the poor Copts who took part in the uprising? Was it merely collective punishment that did not stop to look at who were Muslims and who were Copts? And she knew Ahmed Al-Tayeb's extended family well. From personal experience, she thought that his in-law, Hajjee Sultan from up south in Esna, was a tolerant *shareef*: 'a better man never lived, nor one more liberal to Christians. Copts ate of his bread as freely as Muslims.'[74]

70 See, generally, Ali Barakat, *Tatawur Al-Milkiya Al-zira'iya fi Misr wa Atharahu 'Ala Al-haraka Al-siyasiya, 1813–1914 [The Developments of Agricultural Landownership in Egypt and Their Impact on the Political Movement 1813–1914]*, Cairo: DWQ, [1977] 2017, citing landownership records.

71 El-Sherbini, *Haz Al-Quhuf Fi Sharh Qasidat Ibn Shaduf*, Cairo: Al-Matba'a Al-Amiriya, 1890. See also Gabriel Baer's interpretation of the word *sarraf* ('the Christian tax collector'): Gabriel Baer, *Fellah and Townsmen in the Middle East: Studies in Social History*, London: Frank Cass and Company Limited, 1982, 9, 17.

72 Duff-Gordon, *Lady*, 239–40.

73 Ibid., 229.

74 Ibid.

The Chopping Axe

By now, we know that Shahin Pasha's troops confronted the rebels, with casualties on both sides. The commander might or might not have been wounded. Al-Tayeb was dead or lived on. But, then, Ismail came, and with him Fadhl Pasha, with a much bigger force, estimated at over 10,000 soldiers on board several steamers. According to *Pall Mall*, they were:

> 8,000 Bashi-Bazuks [paramilitary Turkish security forces], and 3,000 infantry, cavalry, and artillery. The troops are stated to have opened with grape and canister on the people crowded together, and then the cavalry was set at them, the Viceroy himself keeping at a safe distance from the scene of this dastardly proceeding. Numbers of the fellahs were killed, and of those caught some were blown from guns by Fadhl Pasha, some hung, and so forth. The remainder, with the women and children, abandoned the villages for the desert. Against these the Maazi and Ababde Arabs [tribal militia from the eastern desert] have been despatched, and we may easily imagine what the deplorable result will be.[75]

From Luxor, Duff-Gordon heard people say that Effandina (the Khedive) have 'taken a broom and swept them clean, [in other words] exterminated the inhabitants'.[76] In a more conservative tone, Nubar Pasha said that Ismail ascended the Nile to Qau 'at the time when executions were being carried out, and the process continued in his presence'.[77] Villages were put under siege, artillery was fired indiscriminately into the masses, and all of the population taken prisoners, Nubar added. On 13 March, Duff-Gordon heard that 100 to 1,000 were already dead, the villages of Qau and Al-'Uqal basin were burned, and women and children were to be sent to other villages. On 12 March, Ahmed Al-Tayeb's brother and his father's father-in-law, Hajjee Sultan, were taken with their families as prisoners to Cairo or Qina, she did not know.[78]

75 'Occasional Notes', *Pall Mall*, 29 March 1865.
76 Duff-Gordon, *Lady*, 215.
77 Nubar, *Muzakirat*, 324.
78 Duff-Gordon, *Lady*, 215–16.

While the massacre in Qau and around it was unfolding, we hear from a British traveller who was on his way north from Aswan that the rebellion was still in full swing south of Qau but steadily being encircled and brought under control. Around 8 March 1865, Rev. Alfred Smith was stranded on his boat at Luxor. The governor told all travellers going downstream that the rebellion ahead at the Girga/Assiout line meant that all boats were by government orders stranded in Luxor until further notice. Smith and other travellers, however, were resolved to move on, since the Nile was rapidly getting lower, which would leave them stranded at Luxor for months before their boat could navigate the river north again. Somehow, they acquired a large amount of guns, rifles, and revolvers, and proceeded north in a small armed fleet bent on facing any potential insurgent action. At Farshoot, they ran into a wounded English engineer who was running Ismail Pasha's Farshoot sugar factory. Coming south from Cairo – where he had gone to procure a new crank for the factory's steam engine – a few days prior, his boat and another were towed by a government steamer. At Qau Al-Kubra, the English engineer saw Ismail Pasha and his troops massacring the population. He met the Viceroy on the shore at Qau and, heeding his self-motivated call of duty, offered to leave him the government steamer tugging his boat, but Ismail declined the offer. Immediately after, five or six miles south of Qau, he faced an effective ambush at a sharp turn of the river that seems to have been put together quickly, where reportedly around 500 rebels, some armed with rifles, came out of palm-concealed shores to fire at the helmsman of the government steamer to stop its advance. Only three rebels boarded the English engineer's boat minutes later, and without harming the sailors, they sought him out and shot him, leaving him wounded; they then abruptly left, as they saw troops advancing towards them from Qau.[79]

Back to Lady Duff-Gordon, on 12 March, her friends the Baron and Baroness Kevenbrinck, who were probably Austrian diplomats, 'saw all the sacking of the villages and said no resistance was offered by the people whom the soldiers shot down as they ran, and they saw the sheep being driven off by the soldiers'.[80] By 15 March:

79 Smith, *The Attractions*, 120–4.

80 Duff-Gordon, *Lady*, 216.

> The Pasha himself we hear is at Girgeh with steamboats and soldiers, and if the slightest fear should arise steamers will be sent up to fetch all the Europeans. [. . .] What I grieve over is the poor villagers whose little property is all confiscated, guilty and innocent alike, and many shot as they ran away. Hajjee Ali tells me privately that he believes the discontent against the Government is very deep and universal and that there will be an outbreak – but not yet.[81]

Duff-Gordon did not agree with Hajjee Ali. She heard about the destruction of whole villages within days. The country, south of Cairo, stretches long and thin, and steamers could come loaded with soldiers wherever a bigger uprising was to break out again, she thought, and kill them all. Maybe there was hope in Cairo and the Cairenes if they rose up? 'Only Cairo can do anything, and everything is done to please the Cairenes at the expense of the fellaheen [. . .] The Cairenes eat the bread and the fellaheen eat the stick.'[82] But what the Lady did not know is that Ismail, before he arrived with wrath and punishment, had deployed 8,000 soldiers in the capital. 'At the same time that this was going on in Upper Egypt, the population of Cairo were in a most excited state; 8,000 soldiers were suddenly but quietly introduced into the city,' reported the *Sun*.[83] By 30 March, Duff-Gordon had a change of heart. She no longer feared the rebels attacking her in Luxor:

> We were all rather looking out for a *razzia* [a raid] and determined to fight Achmet et Tayib and his followers. Then we called them *haramee* – wicked ones – and were rather blood-thirstily disposed towards them and resolved to keep order and protect our property. But now we say *nas messakeen* – poor people – and whisper to each other that God will not forget what the Pasha has done.[84]

She heard a certain Pasha – who was not Fadhl or Shahin – saying that he had hanged 500 rebels; another said that he had sent 300 rebels to join the prisoners taken to Sudan. Perhaps that was upper-class

81 Ibid., 218.
82 Ibid.
83 'The State of Egypt', *Sun*, 15 April 1865.
84 Duff-Gordon, *Lady*, 220.

vigilante action, perhaps just shameless bragging and lies of the victorious class, 'but the most horrid stories are current among the people of the atrocities committed on the wretched villagers by the soldiers'. Rumours were spreading that Ismail had ordered the troops to leave the bodies of the dead unburied for days, and that the stench was horrible. Europeans in the locality were said to have been helping with the massacre. Her close friend Mohamed heard from an Ahmed Al-Berberi that he saw from afar the hats of Europeans surrounding Ismail as His Highness watched the massacre from his steamer in the river.[85] Now she heard that all men, women, and children with any family connection whatsoever to Ahmed Al-Tayeb had been taken prisoners from Salimiya close to Luxor. All were taken in chains to Qina.[86]

On 3 April 1865, Duff-Gordon heard from Sheikh Hassan, a leader of Al-Ababde tribal militia that had aided the government in the massacre – and followed the women and children as they fled into the eastern desert beyond the Qau canal line – that 1,400 men had been beheaded. He saw it happen, as he was taking more of them prisoner to send them to a location unknown.[87] Writing to her friend in Cairo, Mrs Ross, Duff-Gordon wanted to correct her version of the story. In regards to said chopping axe, she said, 'I know very intimately the Sheykh-el-Arab who helped to catch the poor people and also a young Turk who stood by while Fadil Pasha had the men laid down by ten at a time and chopped with pioneers' axes.'[88] The young Turk told her that he wished all peasants of Egypt would get beheaded in the same manner. She heard more about the chopping axes some weeks later:

> Kursheed Agha [a Circassian friend of Lady Duff-Gordon's, a senior officer] came to take final leave before being appointed to Keneh [as Ismail ordered the permanent deployment of troops across Upper Egypt after the massacre]. He had been at Gau and had seen Fadil Pasha sit and make the soldiers lay sixty men down on their backs by ten at a time and chop them to death with the pioneers' axes. He estimated the people killed – men, women, and children at 1,600 – but

85 Ibid., 220–1.
86 Ibid.
87 Ibid., 226.
88 Ibid., 228.

> Mounier [Prince Halim's representative from Muta'na] tells me it was over 2,000. Sheykh Hassan [Ababde militia leader] agreed exactly with Kursheed, only the Arab was full of horror and the Circassian full of exultation.[89]

Death and Vultures

Stranded in Farshoot, south of Qau, Rev. Smith heard in March 1865 that Ismail 'has been taking summary vengeance on the insurgents, hunting them down; burning whole villages'. 'And if [the] report be true,' he added, 'making wholesale slaughter in the disaffected district, and with no little barbarity carrying out his purpose of striking a blow which should intimidate all would-be rebels for the future.' It was, in his opinion, a '*Veni, vidi, vici* with a vengeance'. He heard that once Ismail was in Qau, there was no battle to fight any longer. The thousands of troops were sent out to hang, shoot, and behead hundreds of peasants, before systematically setting the villages on fire.[90] As some 'ring-leaders' escaped to the hilly eastern desert, village sheikhs were ordered to wait for them until they ventured back to the valley from hunger and thirst, and kill them. Then he heard of a further order, 'that if all the rebels denounced were not forthcoming by a certain date, the heads of the sheikhs would go for those they had suffered to escape'.[91]

'The insurgents in Upper Egypt have been defeated' – so went a one-line telegraph from Reuters on 27 March 1865.[92] It must have been received with a sigh of relief across Europe. Two weeks later, the *Sun* was still worried about the European traders in Egypt. The rich Frank traders and entrepreneurs in cotton, sugar, and credit had been saved in Upper Egypt, but what would happen if Cairo rose later? The *Sun* asked:

89 Ibid., 235.

90 Although Rev. Smith mentioned hundreds who were immediately killed on the spot, he agreed with Lady Duff-Gordon's estimates of the 1,500–2,000 casualty toll, adding that she was the most credible source of information available on the uprising: Smith, *The Attractions*, 131.

91 All citations: ibid., 124–7.

92 'The Insurrection in Upper Egypt', *Dublin Evening Mail*, 28 March 1865.

Figure 8.1: Karl Wilhelm Gentz, 'Heads of Rebels'
Ebers, *Picturesque Egypt Vol. II*, 194. © Look and Learn

> Would it not be advisable that England should, as well as France, increase a little the number of her ships of war at Alexandria and Suez, and let the ruling Powers here [in Egypt] know that they would be called strictly to an account for any fatal consequences that might be the result of want of due precaution on their part with regard to the security of the European population of the city [i.e. Cairo]?[93]

Meanwhile, the British press did not report on the axe that felled hundreds of heads. The *Pall Mall* said that 'the majority have been pardoned'. In much more blunt reporting from the battle scene the next morning, it added that 'bodies of the troops have been located in the adjacent villages', and two men accused of killing two Copts were taken to where they had allegedly felled their victims and were beheaded. The

93 'The State of Egypt', *Sun*, 15 April 1865.

same sentence was handed down against two others who killed two Bashi-Bazuks. Then two other rebels, who confessed that they had been soldiers in the past, were shot at the spot.[94]

But Nubar Pasha, who apparently did not intersect with our witnesses on the beheadings from up south, must have laughed bitterly at this report. Cairo streets were already abuzz with tales of the bloodshed. In his autobiography we hear that once Ismail went south to oversee the punishment, the news of the atrocities reached Cairo. Ragheb Bey, a senior official who was close to Ismail and was on his secretarial staff, stormed into Nubar's office to give him the non-British version of the news. He suggested they both send a telegraph to Ismail in the south, 'to tell him about the bad echoes that reached Cairo because of the verdicts carried out in extreme brutality'. It stopped, Nubar heard, but then he asked, 'Was our telegraph the reason or were those still alive from among those wretches too few already?'[95]

In support of Nubar's rhetorical question, our English witness from Farshoot, Mr Smith, as he passed north across Qau on the Nile in his boat, recorded the scenes of the wasteland Ismail left behind in passages worthy of a post-apocalyptic horror film:

> The whole plain of Gow looked deserted and bare. Presently we passed a small village, whose ruined and blackened walls and charred palm-trees told that it had been one of the denounced, and had been fired by the Pasha; while here and there a peasant, stalking with his long gun among the ruins, seemed to be searching among the debris of his home, and looked suspiciously at us as we passed by. But a more ominous sight than all, were the vast flocks of vultures which crowded on the shore. There was the great griffon vulture [. . .] which to the number of above a hundred in a single flock, sat gorged and sleepy, and suggested most repulsive thoughts of the human prey which had attracted them there; even then some were tearing away at a carcase as it floated down the stream. Dogs too were there in unwonted numbers, without masters, sleeping away their cares on the hot sand. And as we sailed on from that horrible spot, all the banks of the river seemed

94 'Troubles in Egypt', *Pall Mall*, 1 April 1865.
95 All citations in paragraph: Nubar, *Muzakirat*, 324.

deserted, which had been so cheerful with men's voices as we ascended the river two months before.[96]

Those who were not killed were reportedly imprisoned and punished. Nubar Pasha heard that 'the remainder of the population were sent to Esna prison, and those who were not imprisoned were banished. After security was regained in full, and the revolution extinguished, the Wali [Ismail] returned to Cairo.'[97]

The Prisoners, the Exiled

Concerning those who were taken prisoner, and those who were sent into exile, we have only even more fragmented pieces of information. Lady Duff-Gordon told her husband on 30 March 1865 that she had recently seen passing close by a steamer taking prisoners to Fazoghlou, in the White Nile, Sudan. She heard that the Ababde militia leader had been ordered to take them across to Sudan through the desert, where river travel was not possible between Egypt and Sudan.[98] On or before 3 April, she sighted a second boatload of prisoners going south, either to Esna or on to Sudan. Hajjee Sultan, Ahmed Al-Tayeb's in-law, was by then known to have been in prison at Qina, and his extended family was imprisoned at Esna, maybe including women and children.[99] A friend of hers, Mohammed, around the same time, boarded another boat that was transporting ninety-eight prisoners, including the *mudir* of Girga, who was in chains and wooden handcuffs, under Khedival punishment for failing to extinguish the uprising in its early days. The Ababde leader was their guard.[100] Then, as late as 29 April, another steamer full of prisoners was spotted going upstream.[101]

By the end of the year, even less concrete news started to proliferate in Upper Egypt about the prisoners and the exiled. It turned into something hovering on legend. In early January 1866, 'Everyone is exasperated – the

96 Smith, *The Attractions*, 129–30.
97 Nubar, *Muzakirat*, 325.
98 Duff-Gordon, *Lady*, 222.
99 Ibid., 229.
100 Ibid., 226.
101 Ibid., 240.

very Hareem talk of the government. It is in the air.'[102] The people were saying that a batch of 100 prisoners sent to Fazoghlou in the White Nile, Sudan, never made it. Their Circassian guard leader, Duff-Gordon had heard, looked at the names of prisoners on his list every morning, as they marched through the desert crossing to Sudan. He would call out names randomly, then tell them they were free to go. Then a short distance into the desert, out of sight, one of his men would strangle them to death. The Lady believed that Hajjee Sultan was treated to the same fate.[103]

Nubar Pasha, meanwhile, reported another story of prisoners that deserves to be cited in full. It was Ismail himself reporting on a certain prisoner from Qau:

> One or two years after these events, one night I was sitting with the Wali [Ismail] at Al-Jazeera palace, after a tour of inspection he conducted in Upper Egypt. He told us that the *mudir* of Esna said to him that an old man in his prison – one of the Qau survivors – talked [with the *mudir*] about the Messiah and his message with strong belief. The prisoner told the *mudir*, 'Can you imagine that we see him a lot entering the hut, then close the door behind him? We then notice smoke coming out from the ceiling of the hut. When we enter, we find that he disappeared, after he turned into that smoke we saw.' The Wali [Ismail] concluded his story by saying, 'I ordered the *mudir* to smash the head of that old man'. The Wali's words were accompanied by a gesture that made our very bones shiver.[104]

Years became decades, and the Qau uprising almost disappeared from memory. In the twenty-first century, it survives only as a far-off, little-known, and fading tale. While the Urabi Revolution of 1881–2 is a defining landmark in the official history of the country, Qau never made it to those records. Among the few surviving traces outside the academic world are two novels, and an anecdote on the Qau prisoners who were sent to Cairo, preserved in writing as it was heard in the 1970s.[105]

102 Ibid., 258.

103 Ibid.

104 Nubar, *Muzakirat*, 325.

105 So far accounts on the Qau uprising in academia that I am aware of are Rieker, 'The Sa'id'; Timothy Mitchell, *Rule of Experts: Egypt, Techno-politics, Modernity*, Berkeley: University of California Press, 2002; Abul-Magd, *Imagined*; Tucker, *Women*,

But while the whereabouts of surviving Qauian rebels and their children form an integral part of the extant oral tradition, less remains known about what motivated their uprising. Apparently, in oral tradition, the ruling class is victorious. The following quote from an oral narrative that survives from the Qau insurrection witnesses this version, as it speaks of Qauians as foolish yet witty robbers of steamers, and no more:[106]

> In the old days, the Qauians used to raid boats carrying wheat in the big sea [the Nile river]. They are dumb people, like the deeply illiterate people of old. Once, they raided a Dahabiya [big sailing boat], and of course they had no idea who was the owner or who were on board. Praise Prophet Mohamed!
>
> Then it turned out that the Dahabiya belonged to the Prince, Khedive Ismail. Troops came and arrested those who raided the boat. The prince ordered their beheading by the Balta [the short axe], and it was done. They beheaded some of them, and sent the rest to the prison at the Citadel – of Mehmed Ali in Cairo.

Conclusion

I started this chapter by proposing that bringing the Qau uprising to life, besides being a worthy act in itself, promises to flesh out some of the consequences of the unprecedented advancement of fossil-powered technology to Egypt with the cotton boom, outside the immediate circle of this proliferation in the Delta. This is what remains to be addressed before we leave the rebels of Qau.

While the people of Qau and the surrounding villages were being exterminated, imprisoned, and displaced, with their villages razed by

---

140; Juan Cole, *Colonialism and Revolution in the Middle East: Social and Cultural Origins of Egypt's Urabi Movement*, Princeton: Princeton University Press, 1993, 196. Cole's two-paragraph treatment of the uprising is unidimensional, reproducing briefly the British diplomats' and the Presbyterian church's narratives. The two novels are Amr Sha'rawi, *Thawrat Qau Al-Kubra* [The Revolution of Qau Al-Kubra], Cairo: Dar Al-Ain, 2019; Maher Mahran, *Qau Usturat Al-dam* [Qau, the Legend of Blood], Cairo: Dar Wa'd, 2010. More on the piece of oral history shortly.

106 Mohamed Abu Zeid Abdallah, from Koum al-Arab (south-west of Qau al-Gharb). Told in the 1970s and retold by Khaled Esmail from memory (and my translation), in Esmail, 'Thuwar'.

Ismail's troops, who travelled on steamers going quickly up the river, the Pasha was in between negotiating his first and second loans from Europe. These loans went largely to importing fossil-powered technology for agrarian production, especially the first loan, and for making the expanding Khedival land possessions – the Daira Saniya – productive of more cotton and sugar. Essentially fossil capital is fossil fuels latched to the means of production, becoming 'self-expanding value passing through the metamorphosis of fossil fuels into CO_2'.[107] In Egypt's case, during the early few years under Ismail Pasha, fossil-fuelled production technology bought on loans was still in the process of being deployed horizontally across agrarian space, along with other modifications and expansions in infrastructures – like the Ibrahimiya canal – that seem to have taken an extremely harsh toll on the landless and small landholders of Middle and Upper Egypt. The Delta was relatively more fortunate with the money that seeped through to the countryside from the cotton sales. Nothing describes the wider context of the Qau spring, and summarises the expansion in the agrarian means of production during the first three years of Ismail's reign, like the prospectus for the second loan, floated to the public in 1866 in London and Paris:

> Under the able rule of the Viceroy, large sums have been expended in a most useful and profitable manner, such as the importation of oxen and horses, the purchase of machinery and implements for irrigation and other agricultural purposes, for the cleansing and pressing of cotton, also the building of stores, warehouses, farms, etc. It is for the further improvement and development of the large possessions of the Viceroy, as well as for the repayment of liabilities already incurred on this account [those of the first loan], that this loan is chiefly required.[108]

The prospectus inviting the public to subscribe to the loan added further that 'the acceptances or obligations of the Daira have constantly been taken by capitalists in preference to all other negotiable securities in Egypt'.[109] The second loan, of £3,387,300 in bonds, was to be issued on

107 Andreas Malm, *Fossil Capital: The Rise of Steam Power and the Roots of Global Warming*, London: Verso, 2016, 290.

108 'Loan of His Highness, Ismail Pasha', *Sunday Gazette*, 25 March 1866.

109 Ibid.

mortgaging 375,000 acres 'from the immense private property of his Highness Ismail Pasha, Viceroy of Egypt'.[110] Then came the 'sugar loan' or 'Daira Saniya' loan of 1870. It amounted to around 13 per cent of Ismail's combined loans from Europe during his years in power and was used overwhelmingly to deploy around twenty of the most advanced sugar-refining factories across the world up to that point. This loan was secured against mortgaging the immense private property of His Highness in Middle and Upper Egypt, the assets so popular to European capitalists when it came to securing precolonial Egyptian loans.

Ismail, in his extreme brutality in Qau and its surroundings, was probably taking a calculated measure to secure and defend, against uprisings, the assets intended for securing the loan taken to develop cash crops in these very regions. This process of developing the cash crops passed through the fusion of fossil-fuelled technology bought on loans from Europe to the land and its people. The geography of the peasants of Qau was one of defiance towards the new arrangement. First it felt the heat of the epizootic, spreading in the Delta after it was imported from Turkey and farther afield, then spilling over to Middle and Upper Egypt. The cattle plague spread because of Ismail Pasha's earliest procedures to plough and irrigate his expanding private land once he ascended the throne, and those of the richest landholders who, like him, were keen on making lucrative profits in cotton after the American Civil War started. Next – or in parallel – came a cascade of mounting taxes in corvée labour and even more animals (camels), to install and operationalise infrastructure for the coming steam engines bought on loans, and to grow more cotton in Sudan by military means. In other words, one can understand the routing of Qau, the axe-massacre across this zone extending between large Khedival and upper-class landholdings, as the solution to a socioeconomic challenge from the subalterns against operationalising steam-powered technology to expand in cash crops across Egypt at the outset of Ismail's reign.

The spring of Qau was not the only type of challenge facing the ambitious Pasha and his plans for sugar and cotton, however. The challenge of expensive coal soon reared its head as well. The grand steam cultivation economy that Max Eyth, chief engineer of Halim Pasha, described in the mid-1860s as already deployed on the Prince's farms – before

110 Ibid.

anywhere else in the world, he added – and which was soon witnessed across Egypt, was soon forced to grapple with the perennial question of coal scarcity. The solution to this problem was to innovate and deploy steam engines feeding on non-fossilised carbons collected from the field. It was a global process, spanning Ukraine in the east to London in the west, with Ismail's Egypt and his private property taking centre stage. We turn to this story, which challenges the straightforward narrative of the deployment of fossil capital, in the next chapter.

9

'Combustible material of every kind'

The Viceroy at the Royal Agricultural Exhibition

Ismail travelled to Europe for several months over the spring and summer of 1867, mainly to attend the International Exposition of Paris, and on this trip he stopped at London.[1] It was Monday, 15 July 1867 when the Khedive left London for the Royal Agricultural Society's Exhibition at Bury St Edmunds, eighty-seven miles north-west of the city. 'His Highness is great in ploughs and harrows,' a local newspaper wrote.[2] The cattle plague had, that year, taken a great toll on England, to the extent that 'no stock was exhibited, and the show, therefore, was, unfortunately, somewhat shorn of its fair ordinary proportions'. The trains employed for bringing visitors to the fairgrounds 'came in one after another with lines of empty carriages'.[3] But then the locals heard of the imminent arrival of the Khedive on one of the trains, and they lined up on the road between the railway station and the exhibition to observe this oriental curiosity, breaking the monotony of the grey day. The Viceroy came with a modest entourage of a few earls and sirs, as well as some 'oriental' attendants. At the show, he never left the carriage.

1 See Timothy Mitchell, *Colonising Egypt*, Berkeley: University of California Press, 1991, Chapter 1, on the Khedive and the 1867 Paris Exhibition.

2 'The Viceroy of Egypt at the Royal Agricultural Show', *Weston-super-Mare Gazette*, 20 July 1867.

3 Ibid.

Perhaps it was too muddy to inspect the exhibited machines on foot. He first stopped at a 7 hp thresher; then he stopped at the stand of Morton & Co. from Liverpool, and asked to see their new cotton-ginning implements, and Morton had on display a model of an iron building for cotton ginning. He stopped at a stone-crushing device that could turn a big rock to small shards in a single stroke and 'soiled' his gloves when he picked up a small fragment to inspect it. He moved on to a little recently patented curiosity: a 1 hp hand-held sheaf-delivery corn reaper, to be used by a man on a horse in the field. His last stop before the Viceroyal carriage returned to the train station was at Howards', where they exhibited a new steam plough to be tried by potential buyers just outside the fairgrounds.[4]

Upon his return to Alexandria in September, he addressed the foreign consuls in Egypt: 'Messieurs – I am very glad to return amongst you. I shall devote my attention to promoting the greatness and prosperity of Egypt. The reception I have met with in Europe assures me of its support in this endeavour. I reckon on the cooperation of all the European representatives in my efforts to develop *the progress of happiness of the country.*'[5]

The next annual fair was held without a Viceroyal visit, but with the progress of the happiness of his country in mind, his own hundreds of thousands of acres included, Ismail sponsored a special prize, outside the usual competitions. A cup of silver and six smaller drinking cups, at the value of £200, were to be granted to a steam cultivator that 'combines strength and usefulness with the least amount of repairs, which cannot be readily executed in foreign parts.'[6] The judges at the exhibition decided to grant it to a cultivator produced by Messrs Fowler & Sons (after John Fowler the father passed away in 1864): a 10 hp steam plough, modified for higher efficiency 'for countries where repairs of machinery are not easily made', that stood at the average cost of £700.[7]

4 Ibid.

5 'Return of the Viceroy of Egypt', *The Tewkesbury Register and Agricultural Gazette*, 5 October 1867. Emphasis added.

6 'The Royal Agricultural Society's Meeting at Leicester', *Bury and Norwich Post*, 21 July 1868.

7 'The Steam Plow in Egypt', *Hillsborough Recorder*, 25 November 1868.

The *Lokombeel*

The news of the Viceroy at the Royal Agricultural Exhibition towards the late 1860s shows that he was not simply a passive consumer of agrarian steam technology. With the cattle murrain of 1863 – which found its way to England by 1866 – he turned to the steam cultivator, the innovative piece of steam technology developed by English engineers early in the decade. The production of the steam cultivator was spurred through his massive purchases by European loans, which supported the expansion and acceleration of the commercial deployment of this technology, at least between 1863 and 1865 (see Chapter 7). His telegram to the Fowler workshops in Leeds in late 1863 is testimony to the most generous Khedival contribution to the development of the steam cultivator, the locomobile at its birthplace, when he offered money to help increase production capacity in the Fowler factory just so it would be able to deliver to his farms 150 steam cultivators by the spring of 1864.[8] The means of producing the locomobile were supported by money from loans taken to buy the locomobiles after the cattle murrain in Egypt (1863). Furthermore, as we might glean from the opening section, Ismail stayed alert and keen on tweaks and innovations that helped customise the locomobile for farms and terrains outside coal-rich regions of the world. As much as he was concerned about expanding the production capacity of this technology (through buying hundreds of locomobiles) and enhancing the technology to fit his farms (through sponsoring a special Khedival prize at the Royal Agricultural Exhibition), further evidence shows active knowledge production from Egypt, over the late 1860s and 1870s, in relation to experimentation with alternative non-fossilised fuels to power steam engines at scale. This process of experimentation on the Nile unfolded in tandem with a larger process of using all sorts of combustible material found on cultivated land – as opposed to underneath it: coal – that seems to have started in coal-deprived Ukraine in the late 1860s.

This chapter, putting the locomobile, or *lokombeel*, as Egyptian engineers and farmers started to name it by the turn of the 1870s, at centre stage, follows this geographically expansive process of innovation.

8 Cited in Chapter 6, Khedival telegram to Fowler of Leeds, 'Quote cost immediate [Fowler] factory expansion. Viceroy willing foot bill.'

Within the space of the 1870s, the locomobile seems to have shifted entirely in Egypt to running on bagasse – sugar cane refuse – cotton stalks, and other combustible organic material on the farm. This investigation, as it happens, brings the mostly forgotten but immense sugar industry of Khedive Ismail to prominence. For it was on sugar cane fields and refineries of the world that modern factories on a massive scale operated in the absence of coal.

The first task before we move to the alternative non-carboniferous fuel is to locate the steam cultivators on the Daira's sugar plantations and factories. How were they deployed and used? What work did they do? One prominent use for the steam cultivators was irrigation from canals and the Nile, to replace the dead cattle and subsequent animal power scarcity. By 1873–5, the Delta alone had around 1,320 steam pumps, the great majority of which were the locomobile type of 5–20 hp range. By this point steam power irrigated 1.5 times more Delta lands than had been irrigated by oxen. In Middle and Upper Egypt, by 1870, the combined horsepower of steam engines deployed in irrigation – outside Khedival sugar cane plantations and factories – amounted to no less than 20 per cent of the combined horsepower of irrigation by steam in 1875 in the Delta: 2,650 hp in Middle and Upper Egypt in 1870 versus circa 14,000 hp in the Delta in 1875.[9] Concerning the Daira Saniya in Middle and Upper Egypt, almost all European visitors there during the last quarter of the century saw many steam engines powering the irrigation of sugar cane, while other engines and some of the same irrigation steam machinery were tied ad hoc to the sugar factories during the sugar production seasons.

In the early 1880s, the British MP Henry Villiers Stuart mounted a horse and toured across some of the then ex-Khedival estates in Middle Egypt. He reported that only the acres under sugar cane – roughly 10 per cent of all land, around 1,000–1,300 acres within no less than 10,000 acres of plantations – were irrigated by steam. 'Irrigation is carried on by a pair of engines of 120 horse-power each', on a 1,250-acre sugar plantation.[10] The rest of the extensive 14,000-acre zone was growing cereals. Charles Dudley Warner gave the same description of cereals-to-sugar-cane ratio

9 See Chapter 7.

10 Henry Villiers Stuart, *Egypt After the War: Being the Narrative of a Tour of Inspection*, London: John Murray, 1883, 322.

at Muta'na in Upper Egypt, but apparently nobody told him about the exclusive use of steam pumps for the sugar cane fields. 'There are ten thousand acres of arable land, of which some fifteen hundred is in sugar-cane, and the rest in grain. The whole is watered by a steam-pump, which sends a vast stream of water inland.'[11] Professor Georg Ebers reported that all sugar cane plantations across Middle Egypt, between Minya and Girga, were irrigated in the mid-1870s by steam pumps, while the cane was tilled and reaped by the hands of peasants.[12]

These machines almost never make an appearance in the Khedival orders under the name 'steam cultivators', but we got the category of the 'locomobiles' (pronounced *lokombeel* as singular, *lokomobilat* plural, in Arabic), probably adapted from French, as the name of choice for steam machinery on the plantation.[13] These locomobiles must have been the engines seen by Bayard Taylor, who commented in 1874 on this mobility, 'The scarcity of animal power led to the introduction of small portable steam engines for pumping water.'[14] By the 1870s, the *Sugar Cane Magazine*, the trade journal that provided sugar capitalists with expert knowledge, stated in passing that 'engines of the "portable" class [are] now so largely used in all agricultural operations [i.e. across the globe]'. We can safely deduce then that the locomobiles in the official state records are those portable engines, another common name for the steam cultivator.

The portable engines were handling several tasks in Egypt hinging on their mobility, hence the naming locomobile. At Al-Samata, Girga, MP Stuart saw a steam-pumping engine with a tall chimney. It was

11 Charles Dudley Warner, *My Winter on the Nile*, Boston: Houghton, Mifflin and Company, 1876, 354–5.

12 G. Ebers, *Egypt: Descriptive, Historical and Picturesque, Vol. II*, translated by Clara Bill, London, Paris, and New York: Cassell, Peter Galpin & Co., 1879, 160.

13 Khedive Ismail was literate in French. Also, French words for machines and machine parts are common in Egypt even in the present, for example *dereksion* (*direction*) for the steering wheel, *fetais* (*vitesse*) for the gearbox, *marsheder* (*marche arrierè*) for moving the vehicle backwards, *wabour* (*vapeur*) for steamers and many agricultural mechanical prime movers, and so on. A mention of 'steam machines known as "locomobiles" provided by the government' in Amin Sami, *Taqwim Al-Nil 3.2 [Survey of the Nile Vol. 3.2]*, Cairo: Matba'at Dar Al Kutub, 1936, 793. An inventory of possessions of a certain estate close to Cairo included 'locomobiles' as well as equipment and cattle; see Khedival order to Daira Saniya, 29 Jumada Awal 1286 H [6 September 1869], in ibid., 827.

14 Bayard Taylor, *Egypt and Iceland in the year 1874*, New York: G. P. Putnam's Sons, 1879, 137.

the property of 'a wealthy Pasha'. It fed a 100-acre piece of sugar cane on a 1,000-acre estate.[15] The estate manager told him that outside the irrigation season, when it was time for the sugar manufacturing, the same engine was moved and used at the sugar factory.[16] We do not know, however, if this was a small mill owned by the same Pasha, or if the engine was hired temporarily by the nearest Khedival sugar factory. Here, the mobility of the locomobiles was essential for their flexible powering of various operations, depending on which season it was and what kinds of farm and mill work the engines were required to power.

Another merit that Stuart reported regarding the Pasha's locomobile is that it did not run on expensive coal. Its furnace was fed with 'the straw of sugar-cane', the bagasse, and complemented with dourra (maize) straw.[17] Another larger estate, of 16,000 acres, owned by a wealthier Ahmed Pasha, had its 120 hp steam pump – probably a Cornish pump – irrigating the 800-acre sugar cane plantation fed by 'cane straw grown on the estate, and with dourra-straw purchased at the market'.[18]

This was a central feature in the fuel economy of agricultural steam machinery across Egypt in the 1870s, including in the sugar factories. In fact, Stuart reported on the steam cultivators' fuelling system across Egypt in the early 1880s as being 'independent of coal, as the furnaces are fed entirely with straw of maize, dourra, and sugar-cane. I have seen 120 horse-power engines thus fed.'[19] The most modern sugar production establishment was no different. A decade earlier, in 1872, William Mein, the chief engineer of the Khedival sugar factory at Rhoda, told the *Sugar Cane Magazine* that all fuel for the machinery and furnaces under his watch was 'megass', or bagasse, the dried crushed shoot of sugar cane after all the juice is squeezed out of it.[20]

15 Stuart, *Egypt After*, 260.
16 Ibid., 261.
17 Ibid., 260.
18 Ibid., 263.
19 Ibid., 242.
20 *Sugar Cane Magazine*, 1 October 1872, Vol. IV, 550.

The Forest, the Bagasse, the Soil

It was no coincidence that sugar factories in Egypt were early sites for fuelling the steam engines by bagasse. This practice belongs to a tradition that had started centuries before the invention of the steam engine itself.

While the steam cultivator's use of biomass – in the absence of the coal that we commonly associate with steam engines – seems counterintuitive, biomass on the sugar plantation was the essential fuel for sugar production centuries before the Industrial Revolution, at least as far back as the fifteenth century. Sugar refineries were built in northern Europe to process unrefined sugar imported from elsewhere, mainly because of the abundance of dense forests providing firewood for the boilers. In seventeenth-century Brazil, a large Bahian sugar manufactory consumed 12,000 to 13,000 pounds of firewood per day, during the harvest and refining season. An acre of sugar cane consumed the trees of 1.5 to 2 acres of forest. In Barbados, where serious sugar production started in the mid-1640s, all forest was annihilated by 1667, and the sugar refiners there then had to import coal for their boilers from England. The sixteenth-century Mexican sugarworks seriously harmed the indigenous groups' firewood supplies.[21]

Under the pressure of deforestation, a transition eventually happened to bagasse across the nineteenth century. Earlier, manufacturers did not consider the cane trash a serious alternative to firewood because, in the early modern period, cane-crushing wheels at mills did not apply enough pressure that could by-produce bagasse dry enough to burn in boilers.[22] If the overly damp cane refuse back then was left to dry in the shade until it was fit to use as fuel, its high level of lingering moisture meant it would rot on the sugar left in its fibres before it was dry. If less damp at the crushing mill, with more efficient cane-juice extraction squeezing more liquid out of the shoots, it could be dry before it rotted and hence became usable in the furnaces as fuel. With the development in the efficiency of the cane-crushers across the nineteenth century,

21 Multiple sources cited by: Jason W. Moore, 'Sugar and the Expansion of the Early Modern World-Economy: Commodity Frontiers, Ecological Transformation, and Industrialization', *Review (Fernand Braudel Center)* 23:3, 2000, 409–33, 422–3.

22 Ibid., 423.

propelled partially by seeking higher productivity and partially to use the bagasse as fuel when deforestation under the influence of sugar devastated the nearby woods, came the shift to bagasse: in Cuba around mid-century, in Brazil some decades earlier, and in Peru through the 1870s.[23]

After centuries during which the sugar industry annihilated the forests, it became so common by the early 1880s to use bagasse as fuel in the industry that a treatise on sugar production from the decade stated that

> the sugar factory consumes a very large quantity of fuel, and fuel in the shape of coal or wood is usually very dear and scarce in sugar growing districts. Hence has arisen the *generally-accepted custom* of using the begass for fuel.[24]

If, by the early 1880s, it became a generally accepted custom to use bagasse for fuel in the sugar factories, that becoming needs to be accounted for. A good point of departure for this account is to return to Egypt and look for coal in the sugar factories before unfossilised biomass became the only fuel reportedly powering the works, being alternative to the prohibitively expensive coal. From there we can trace how bagasse and other plant matter replaced coal. Another no less promising point of departure is to explore the zeitgeist of sugar factory fuel technology in the discussions of the sugar capitalists of the world before the late 1870s, especially among those in other coal-deprived regions across the globe. But since we are almost sure by now that, in the 1860s–70s, global agricultural mechanisation endeavours unfolded on the banks of the Nile no later than they did anywhere else, and the development and deployment of the British steam cultivator is our witness, we can safely travel both roads together and see where this takes us.

The chief engineer of the Rhoda sugar factory, Mr William Mein, who in 1872 wrote to the *Sugar Cane Magazine* and said that all fuel in his factory was bagasse, said the same to a prominent European visitor

23 Several sources in ibid., 423.

24 C. G. Lock, G. Warnford, W. Winger, and R. H. Harland, *Sugar Growing and Refining: A Comprehensive Treatise on the Culture of Sugar-Yielding Plants, and the Manufacture, Refining, and Analysis of Cane, Beet, Maple, Melon, Milk, Palm, Sorghum, and Starch Sugars*, London and New York: E. & F. N. Spon, 1885, 38. Emphasis added.

to his factory around November 1869.[25] Nonetheless, coal was being used in Egyptian sugar factories in 1869 and earlier. The visitor in question wrote a letter to the same magazine accounting for his visit to the Rhoda factory upon invitation from Ismail himself. At Rhoda, he met the chief engineer, who said they ran the factory on bagasse. Elsewhere in his letter, as he commented on the state of sugar production in Egypt generally, the same prominent guest of the Khedive said:

> I was told that besides the straw and megass [bagasse], they burned [across Egypt] every year 70,000 cantars [12,300 tons] of coal for making about 100,000 cantars [17,568 tons] of white-looking sugar, and that coal costs in Rhoda at least four times as much as it would cost a refiner in Liverpool.[26]

So the Rhoda factory was not running on coal at all, but the sugar industry consumed an annual average of 12,300 tons of coal. Other factories and production processes outside the factories must have been consuming all that coal, whose weight stood at 70 per cent of that of the sugar it helped produce.

The consumption patterns of all this coal as production inputs for the sugar, for our purposes here, should not detain us for long. It could have been consumed inside the furnaces as an auxiliary combustion agent with bagasse, for crushing cane, in boiling syrup, and rotating centrifugal machines of many other factories beside the one at Rhoda. A substantial part of that coal could have been counted as related to sugar because it ran steamers and/or light railways that moved between plantation and factory, or between depots of other production inputs – including steamers that towed the boats bringing the labourers – and factory. What we know more about are certain indicators on the broad contours of the shift in combustibles inside the engines' fireboxes towards total unfossilised biomass in the form of bagasse, cotton stalks, and other kinds of straw, in Egypt and elsewhere.

Starting from elsewhere, we get the impression, at the turn of the 1870s, that sugar capitalists were relying on bagasse for fertilising the

25 *Sugar Cane Magazine*, 1 October 1872, Vol. IV, 550, and 1 July 1870, Vol. II, 387.

26 Ibid.

sugar fields after harvest, and that, simultaneously, the use of bagasse as fuel for the sugar production was a known practice, conducted with a guilty reluctance, given the competition with fertilising. How was bagasse applied to the sugar machinery? A member of the Royal Institution of Civil Engineers provided a detailed description from Mauritius:

> Mr. Longridge said [. . .] he had seen in Mauritius a good deal of the use of 'bagass,' as it was there called. In some of the best sugar estates this and cane straw were the only fuel employed for the [. . .] mill, the turbines, and for cooking the sugar. The bagass used was generally that of the preceding crop. It was stored in large stacks, roofed over, and kept till the next year. If taken as it came from the mill, it contained too much moisture. Formerly the bagass had to be supplemented by the use of coal or wood, but he could not state the proportions. There was no difficulty in the feeding. There was a large vault, almost like a baker's oven, into which the men threw the bagass; one man fed the fire with a shovel, throwing the fuel pretty uniformly over. There was a good draught, and capital combustion was obtained. The best mills had tubular boilers, and they not only got up the steam for the engines, but cooked the sugar without other fuel.[27]

What was left from the bagasse, originally useful as fertiliser, was nothing more than its ashes. And those ashes were the focus of a heated debate over the first few issues of the *Sugar Cane*.

The second number of the *Sugar Cane* came out in September 1869, on its cover page a long essay from the renowned chemist Robert Angus Smith, a prominent analytical chemist in Her Majesty's government. Dr Smith discovered acid rain in the 1850s by investigating pollution in northern British cities – Manchester featured prominently – and was later dubbed 'the father of acid rain'. He published *Air and Rain: The Beginnings of a Chemical Climatology* in 1872, becoming a pioneer of climate science.[28] His article for the *Sugar Cane* started with the

27 *Minutes of Proceedings of the Institution of Civil Engineers*, 1876–77, Vol. 48, London: The Institution of Civil Engineers, 96–7.

28 Robert Angus Smith, *Air and Rain: The Beginnings of a Chemical Climatology*, London: Longmans, Green, and Co., 1872.

chemical analysis of two samples of bagasse ash, collected from 'a large and powerful mill in the West Indies'.[29] Then a section on bagasse as fertiliser, and another on bagasse as fuel; an expert intervening in a heated debate among practical and less scientifically informed men. For the bagasse was apparently being burned for the sugar factories for quite some time by then, and producers were debating whether or not the ashes left over from the burning would fertilise the land after harvest as well as the intact bagasse. He gave his judgement on manuring, saying that the burned ashes 'possess largely those mineral matters upon which the plant has fed', albeit 'in a more concentrated form'. From his chemical analysis, the bagasse ash makes 'a valuable substance', as it 'contains pretty nearly all the mineral matter necessary for a future crop'. Notwithstanding this reassuring result, the soil, readers should know, lost more than just the minerals, making the bagasse ashes a less than perfect fertiliser that is still very useful for the fast-deteriorating soil under sugar cane.[30] After establishing the facts on the usability of burned bagasse, he moved to its usefulness in burning as fuel. His chemical analysis on bagasse fuel amounted to the conclusion that 'Wigan Cannel coal gives, therefore, on complete combustion 2.36 times more heat than megass, and requires 2.18 times more air to burn it'.[31]

From Antigua, a certain Mr W. T. Blackburn, in the October 1869 issue, volunteered his opinion on the trade-off between bagasse as manure versus fuel. He argued for spending only half the money usually spent on buying foreign fertilisers, to purchase enough coal for sugar processing, while using the freshly crushed bagasse as fertiliser, for it kept the land hidden from the scorching sun in the dry seasons, 'whilst the slow decomposition of the megass would restore nearly all that had been taken from the soil'.[32] As good coal that burned intensely to white ash was available for shipping from Liverpool or Manchester to Antigua, why burn bagasse when you could use it as fertiliser and save money? From St Helena, December 1869, a letter to the editor from a sugar producer informed the readers that land in the West Indies was exhausted, as bagasse had been burning there as fuel for too long. Native

29 *Sugar Cane Magazine*, 1 September 1869, Vol. I, 68.

30 Ibid., 67–8.

31 Ibid., 70.

32 *Sugar Cane Magazine*, 1 October 1869, Vol. I, 173.

growers and manufacturers should learn the lesson. 'With a little care at the commencement, and the substitution of coal or wood for fuel [instead of bagasse] so as to allow all the trash and megass to go for manure, surely the soil can be retained [. . .] prolific'.[33]

From the colony of British Honduras (now Belize), a gentleman who visited the works in passing reported his first expert impressions. He was taken aback by the 'too great a waste on all the estates': this country abundant in wood fuel should have consumed the forest and returned the bagasse to the soil, right after the cane was crushed at the factory, 'so as to compensate for the substances withdrawn [from the soil] in the sugar'.[34] February 1870: another gentleman, whose father owned a sugar estate in the Demerara colony (now part of Guyana), advocated for returning the burned bagasse ash to the soil, not the fresh form. He took it for granted that bagasse was destined for powering the factory machinery. He reminded those who used the bagasse ash as a component in manufacturing bricks – something that Khedive Ismail did – that the ashes were destined for the role of fertiliser, applied in a finely ground state, better if combined with a mix of ammonia sulphate and lime phosphate, or Peruvian guano.[35] A piece from April 1870, covering a public lecture held in Edinburgh on sugar manufacturing, seconded this opinion on using bagasse ash for manuring, and marvelled at the lost opportunity in Demerara.[36]

But then strong opposition to bagasse ash only as fertiliser came from another analytical chemist, Dr Thomas Phipson, member of the Chemical Society of Paris. He categorically refused the use of bagasse ash because 'we cannot expect to get work twice out of the same material'.[37] The ammonia sulphate that our gentleman from Demerara advised fellow sugar growers about, according to Dr Phipson, had the effect of healthy-looking cane shoots, rich in water and poor in sugar, and highly exhaustive to the soil.[38] In May 1870, Dr Phipson's argument against using the burned bagasse as fertiliser received a harsh comment from Mr Manley Hopwood. His judgement on Phipson's unpractical

33 *Sugar Cane Magazine*, 1 December 1870, Vol. II, 677.
34 *Sugar Cane Magazine*, 1 January 1870, Vol. II, 21.
35 *Sugar Cane Magazine*, 1 February 1870, Vol. II, 96.
36 *Sugar Cane Magazine*, 1 April 1870, Vol. II, 215–16.
37 *Sugar Cane Magazine*, 1 March 1870, Vol. II, 142.
38 Ibid.

science was: 'if the grower can find other fuel to hand, and as cheap as cane trash or megass, it may be well for him to consider Dr Phipson's proposition'.[39]

Around the time this debate was raging, we get a lengthy report on the findings of a commission at Réunion, which inquired into the deteriorating state of the cultivated sugar cane there. The commission said that the deteriorating state of soil under the diseased cane of Réunion was because bagasse had not been returned to the soil for too long. Sugar cane was grown in the same land for long seasons without intermissions. Guano was applied for manuring, but not applied correctly.[40] Neighbouring Mauritius, which also suffered deteriorating soil and diseased crops, recently received gratis samples of new cane species from Egypt and Queensland, to try to overcome the problem. Concerning Réunion, Mr L. Wray, a member of the commission, lashed out against bagasse fuel, given the lost opportunity of applying it as fertiliser: 'using the cane-trash as fuel is the most unwise and suicidal practice I can imagine; and I believe no planter will continue such a system who can possibly obtain a supply of wood or coal'.[41]

The commission also lashed out against M. George Ville – the agronomist from Napoleon III's imperial gardens, who marvelled at how the light of the steam cultivator entered the fellah's soul in Egypt a few years earlier – who was advocating for dry chemical fertilisers at the time, instead of applying green and moist plant manures to the soil. Finally, the commission recommended not to ever overlook 'the necessity of returning to the soil, all the elements of nutrition of which it is deprived by the cane', to apply bagasse evenly, and other natural manures, including the molasses by-produced with the white sugar in the factory.[42]

Such was the state of the debate among the practical men of sugar production at the turn of the 1870s: a complicated node of 'trade-offs' between using bagasse as fuel or fertiliser. If coal or wood is readily available, do not buy fertilisers and apply bagasse. If both is expensive, burn the bagasse as fuel and apply its ashes as fertiliser and deplete the

39 *Sugar Cane Magazine*, 1 May 1870, Vol. II, 295.
40 *Sugar Cane Magazine*, 1 June 1870, Vol. II, 305–6.
41 Ibid., footnote in page 310.
42 Ibid., 314.

non-renewable soil as the ashes do not carry the same benefit of nitrogen-rich and moist cane trash back to the land. Meanwhile, news was coming out already that the process of consuming bagasse as fuel was wreaking havoc on soil under cane, and everything was breathing in a strong air of uncertainty and covert denial. However, one can recognise a certain steady trend towards soil depletion with the strong push for using the cheap bagasse as fuel. But if the soil suffered under the capitalist's feet, that was a concern for the future. The profits of the current harvest topped the agenda, and the surplus would increase as long as the planter did not pay for imported coal or fancy natural fertilisers. Forests gone up in flames, gobbled up by the sugar factories, but, thankfully, bagasse was now here to burn: all from the cheap surrounding nature.

However, by 1873–4, the relatively long-term influence of consuming bagasse as fuel for sugar production did show through in hard numbers on sugar cane productivity at major production centres across the globe:

> It is remarkable that the production of exotic sugar [cane sugar] has in 20 years only increased from 1,276,714 tons to 1,811,826 tons, and also that this increase is confined to a few countries. The chief of those in which progress has been made are Java, Cuba, Manilla, the French Antilles, Jamaica, English Guinea, *and Egypt*. Everywhere else the production of cane-sugar has extended but little, and in several cases it is notoriously on the decline. The great colony of Cuba itself only shows slight progress.[43]

Ismail could have been following this debate, perhaps by means of briefings from his senior engineers. The magazine found its way to Mr Mein, the chief engineer of the great Rhoda factory. We know that because Mr Mein wrote a letter to the editor in 1871 or 1872, a couple of years after the magazine was founded. Even if Ismail was not following the *Sugar Cane*, 1870 was the year he took the great 'sugar loan' from European banks, and it was the time of intensive sugar-factory building across the *Daira*. Fuelling these factories must have been a central concern of his, and we should expect that the Rhoda factory model – running fully on bagasse – was one he looked to copy for the factories to come.

43 *Sugar Cane Magazine*, 2 March 1874, Vol. II, 106. Emphasis added.

In fact, we know that he did more than follow the debate. He was actively testing the efficiency of burning bagasse and other local agricultural trash for generating steam inside his locomobiles and factory boilers designed and built to burn coal. Under Ismail's watch, the testing and usage of alternative fuels for the steam engines and boilers did not stop at the purpose of producing sugar, for it extended across agricultural operations of various kinds, and innovation came first from the locomobiles before it was transferred to the fixed large boilers of the sugar factories. This story reportedly started from the grain fields of Ukraine, probably around the time sugar capitalists were debating the burning of bagasse under pressure from deteriorating soil conditions.

The Head-Schemioth Steam Engine

Like Ismail in Egypt, we are told that once the portable steam engine found its way to the Russian Empire, the average farmer started looking frantically for alternative ways to fire the engine, to thresh the grain.[44] The empire's grain farmers, like their counterparts on the Nile, were facing the same challenges of costly coal imports, and the scarcity of firewood:

> Those thinly-populated stretches of flat are the very places for the profitable employment of steam, but then, unluckily, there are fewer trees than men, and even scrub and brushwood are often wanting. As for coal, its price, as laid down at the railway stations, is almost prohibitory, and even with the new extension of the railways you might often have to haul it great distances.[45]

Whether steam-powered grain threshing was the motive is a question we will not pursue, but it was indeed an engineer from Odessa, Mr Casimir Schemioth, who reached out to a British portable engines manufacturer in the early 1870s, to cooperate on building a portable steam engine running on straw and other agricultural biomass.[46]

44 'The Flemish Farm', *Evening Mail*, 18 May 1874.

45 Ibid.

46 'Straw Burning Engine', *Press Monday* (New Zealand), 23 February 1874. Schemioth's full name occurred in the British Patents Office's guide for the year 1872. See Bennet Woodcroft, *Chronological and Descriptive Index of Patents Applied for and*

Schemioth must have been actively involved in the early Russian experiments for feeding steam engines straw to thresh grain. He might have been the very first pioneer, or one of many others we are unable to meet through the archives. The narrative on the early Russian experiments follows the same logic of the transition from the wood and coal to bagasse for sugar production. Here, as the *Evening Mail* reported, it was a transition from both to straw, for the purpose of grain threshing:

> The superfluous straw is a positive embarrassment to the farmers. It is stacked on the steppes and left to rot; very little of it is used for manure, for the soil is naturally excessively rich, and cultivated for the most part on the system of a single crop following two years of fallow. Casting about for a substitute for ordinary fuel, it occurred to the Russians that straw might serve their purpose, and they did succeed in applying it in a clumsy and costly fashion. They dug great pits constructed with a side funnel; they brought the engine bog over the mouth of the funnel, and they did their threshing *tant bien que mal*. But the process was too expensive to come much into favour. The valuable engines were used up quickly; the grain had to be brought to the pit, and there was a great waste of heat with an immense consumption of straw.[47]

What would we do if we were in Mr Schemioth's shoes, standing there watching the imperfect process of igniting the engine with heat applied from below? Such a waste of straw transported to the pit under the engine, by means of much animal and human energy! Straw stacked in dense heaps did not have enough air flow to keep an efficient fire – an imperfect setup that did injustice to the modern self-acting machine above the pit. The engine became charred, blackened, and wasted before long. Moreover, the wheat had to be transported to the then crippled portable engine, which lost its advantage of mobility once it became tied to the combustion station.

As a remedy, we might think of a way to burn the straw inside the engine's firebox, not underneath it. And who were producing excellent

Patents Granted (London): Office of the Commissioners of Patents for Inventions, 1873, 22.

47 'The Flemish Farm', *Evening Mail*, 18 May 1874.

portable engines for agrarian production at the turn of the 1870s? Messrs Fowler of course, and Messrs Howard, at least during the 1860s, but there was now, and since the late 1860s, the firm of Ransomes, Sims, and Head of Ipswich and London. Just before 1870, they won, according to their own advertisements, several prominent industrial prizes across Europe – including the First Grand Prize and Gold Medal of the Paris Universal Exhibition of 1867 – India, and South America.[48]

Perhaps Mr Schemioth wrote to them and others, or more actively reached out and frequented the exhibitions of western Europe and the agricultural fairs of the even more advanced England. He finally met John Head in or before 1872. For, during that year, experiments started in earnest on putting the straw inside the engine. According to Mr Head, who presented a lengthy paper about the Head-Schemioth invention at the Royal Institution of Civil Engineers, they experimented with

> a locomotive boiler adapted to a portable engine, in order to perfect the combustion of straw and other vegetable refuse. It was decided to inject the fuel automatically into the furnace, by an apparatus not unlike the feed motion of a chaff-cutter. In the construction of the boilers which resulted from these experiments, the necessity of adapting them to the consumption of almost every known species of fuel was constantly kept in view, and the engines manufactured on Head and Schemioth's system will burn not only the refuse vegetable substances already enumerated, but also coal and wood, peat of all kinds, sawdust, chips, and other similar materials.[49]

Before January 1872, Schemioth had already moved to England. He applied there for a patent, and it was granted on 11 January 1872 exclusively to the name of Casimir Schemioth of Odessa, residing in Middlesex county, without any mention of the engineers of Ipswich.[50] The entry describing the invention was short: 'Improvements in steam boiler and other furnaces, and in parts applicable thereto to facilitate the combustion of straw or other light matters as fuel.'[51] But when the invention was

48 *Grace's Guide to British Industrial History* website, page 'Ransomes, Sims and Head' (accessed 1 March 2026).

49 *Minutes of Proceedings*, 1876–7, Vol. 48, 76–7.

50 Woodcroft, *Chronological and Descriptive*, 22.

51 Ibid.

registered as a trademark in the United States in November 1873, the name of John Head came up as an associate, and the registration entry indicated that Schemioth was now (by late 1873) deceased.[52] Perhaps the Russian engineer patented his blueprints of the invention before he started experiments with Head, or perhaps they both agreed that the former should be the holder of the patent since the latter merely cooperated in realising the patent's simple but effective innovation. Or perhaps he did not even actively participate in the conversion of his blueprints into a working engine after patenting the method.[53] In any case, Schemioth died before the Vienna Exhibition that started in May 1873, where the fruits of his collaboration with the Ipswich engineers posthumously shined bright.[54]

From Vienna, the *Engineering* weekly journal provided a thorough coverage for the straw-burning portable engine. One of the three illustrations it provided (Figure 9.1, bottom) shows the furnace of the portable engine, attached to a tray on which straw and other combustible material from the field are moved forward to feed the furnace. At the mouth of the furnace is Mr Schemioth's innovative tool (two small cogs at the end of tray; Figure 9.1, top). It contained a small mechanism: two rolling fans that could be rotated by steam power – as it was hooked to a belt attached to the engine's parts moved by steam – or through a hand tool to be fixed at the end of the tray to manually roll in the straw to the middle of the furnace, not from the top.[55] The manual tool allowing rotational motion was for the first ignition, to generate enough steam that could later take up the task of stoking by moving the tray.

52 Patented by William Westlake (administrator of now deceased Casimir Schemioth) and assignor to John Head, Ipswich, England. Patent application filed 6 August 1873. See *Official Gazette of the United States Patent Office, Vol. IV*, 1 July–30 December 1873, Washington DC: Government Printing Office, 1874, 483 (patent no. 144,241).

53 'Straw Burning Engine', *Press Monday* (New Zealand), 23 February 1874: 'But although [Schemioth] was ultimately on the way to comparative success, he left the Messrs Ransomes, Sims, and Head, whose cooperative assistance he solicited, practically to solve the problem.'

54 'The Vienna Exhibition', *London Evening Standard*, 9 May 1873: 'Mr. Head and Mr. Schemioth, a Russian Engineer, *now deceased*, put their wits to work, and this engine is the result.' Emphasis added.

55 *Engineering*, 23 May 1873, Vol. 15, 358.

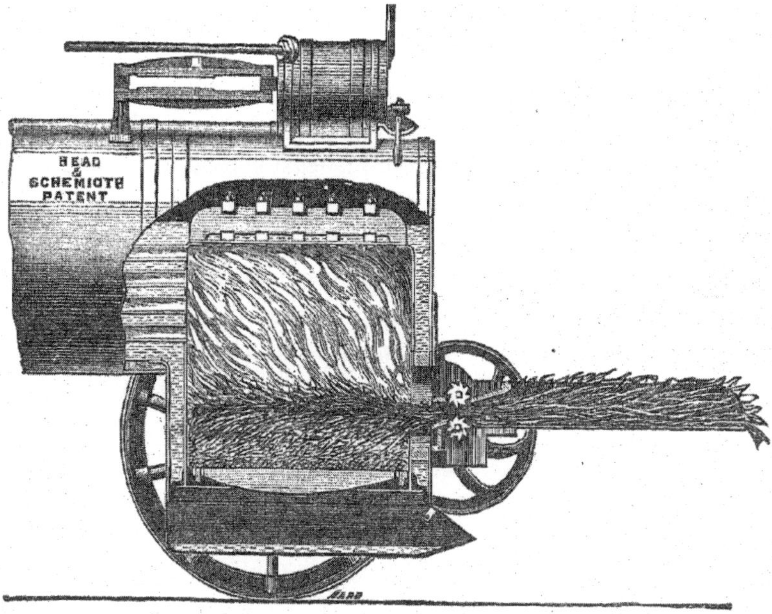

SECTION THROUGH FIRE-BOX.

Showing position of Straw or other Vegetable Substances in process of combustion.

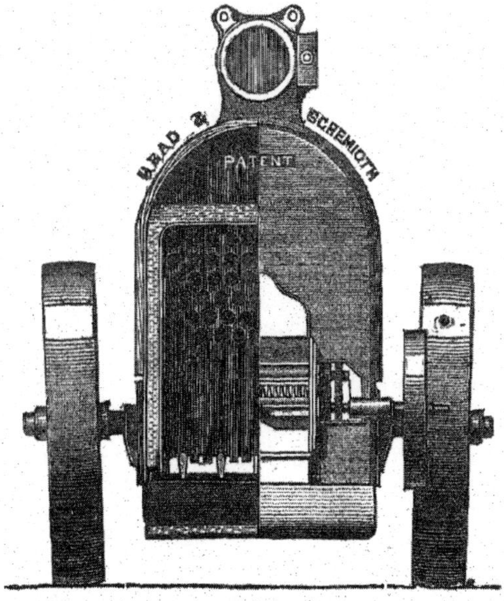

END VIEW OF ROLLERS AND FIRE-BOX.

Figure 9.1: Design of Head-Schemith improved portable engine

The ingenuity of Schemioth's patent, according to the *Sugar Cane*,

> is that by means of a continuous mechanical feed, the fuel can be forced into the furnace in a thin stream in the form of a fan, and the fresh fuel is practically held in suspension for a short time [i.e. flies inside the furnace before it lands slowly on the burning heap], allowing the separate stalks to become immersed in the flames, and the long pieces of straw, reeds, or brushwood, to have the effect of stirring up the half-burnt material in the furnace, thus keeping the whole in motion besides permitting a proper ingress of atmospheric air, which is necessary for the rapid combustion of vegetable matter.[56]

The *Engineering* journal documented a trial of the engine at Ipswich in mid-May 1873. The successful test results reached as far as the farmers of New Zealand, where the *Waikato Times* copied the story. The first ignition, it was noted, was easy enough to be 'performed by a boy'. In under an hour, the pressure reached 60–70 lbs, producing '20 effective horse power [. . .] the whole of the arrangements work[ed] admirably', and the straw's combustion was 'thorough and complete [. . .] and altogether the trial was a most satisfactory one in all respects'.[57]

At the Vienna Exhibition, Dr William Siemens, the younger brother of the founder of the Siemens company, was among the judges.[58] He later told his fellows of the Royal Institute of Civil Engineers Institute that the jury was undecided between the Head-Schemioth engine and another presented by a Mr Garrett. 'The jury, he believed, decided that both makers were entitled to an award', but he selected the first to be granted the Diploma of Honour, because it could be operated smoothly and systematically by unskilled workers, whereas Garrett's engine needed skilled hands to avoid its tendency to explode.[59] The award-winning engine garnered great attention, it seems. We are told that it was 'one of the great centres of attraction in the agricultural machinery

56 *Sugar Cane Magazine*, 1 January 1878, Vol. X, 31–8.

57 *Engineering*, 23 May 1873, Vol. 15, 358; copied in 'Facts for Farmers: Straw-burning Portable Engine', *Waikato Times* (New Zealand), 13 September 1873. Emphasis added.

58 Siemens global website.

59 *Minutes of Proceedings*, 1876–7, Vol. 48, 99.

Figure 9.2: Illustration of the Head-Schemioth improved portable engine

hall [at the Vienna Exhibition]', and by May 1874 it had 'already obtained the approval of the most competent agricultural authorities'.[60]

The patent was adopted in the mid-1870s by Messrs Fowler to their line of pioneering steam ploughs, which became 'supplemented by means of supplying vegetable fuel with a fork by hand, through an opening in the fire-box, fitted with a balance door, which closed immediately the fuel was inserted'.[61] The Ipswich engineers advertised their invention around the

60 'The Czar's Steam Thrasher', *West Sussex Journal*, 19 May 1874.

61 John Head, 'The Combustion of Refuse Vegetable Substances', *Irish Builder: Architectural, Archaeological, Engineering, Sanitary, Arts, and Handicrafts* 19, 15 February 1877, 60.

same time, as being already 'tested for some time in Russia and on the banks of the Danube, *in Egypt* and India, and hav[ing] been found to effect a great saving in fuel, enabling steam power to be introduced into countries where it was formerly impossible to make use of it'.[62]

Burning Combustible Material of All Kinds

The essay John Head published in *The Irish Builder* on the invention some years later, and his paper presented at the Royal Institution of Civil Engineers, expanded on the Egyptian connection. Right after the great success at Vienna, Ismail Pasha commissioned the Ipswich engineers to supply a modified apparatus, 'for burning cotton stalks [in] a tubular boiler, constructed by Messrs. J. and H. Gwynne, for supplying steam to a 12 nominal hp pumping-engine'.[63] From his presentation at the Royal Institution of Civil Engineers, on 30 January 1877, Mr Head added that Egyptian cotton stalks after they are pulled from the soil after harvest – 0.5 inches in diameter, and 4–5 feet in length – burned as efficiently as good wood.[64] But, even before Egypt adapted that 12 hp irrigation pump to run on cotton stalks, at Cairo, in March 1874, cotton wood fuel was tested in a portable engine of 10 hp, giving the ratio of 1 to 2.98 in combustion per weight unit (coal to cotton stalks).[65] To put this in perspective, it was reported that the cotton stalks of one acre in the Nile Delta became sufficient to generate steam in the engine running to irrigate that same acre.[66] At the Institution, Mr Head added:

> No doubt the refuse of cotton plantations can be largely used as fuel for steam boilers in all districts where cotton is grown [in Egypt]. [. . .] Several portable engines for pumping are now [1877] at work in Egypt, in which cotton stalks alone are used as fuel, and in which steam is kept up with the greatest regularity. Ultimately it will

62 'Advertisement by Ransomes, Sims & Head, Orwell Works, Ipswich', *London and China Express*, 10 July 1874. Emphasis added.

63 Head, 'The Combustion', 60.

64 *Minutes of Proceedings*, 1876–7, Vol. 48, 76.

65 Ibid., 86.

66 *Sugar Cane Magazine*, 1 January 1878, Vol. X, 38.

> unquestionably form one of the most important articles of fuel in Egypt, where wood is not indigenous, and coal also has to be imported.[67]

We know from the *Sugar Cane* debate covered previously that Middle and Upper Egypt were already relying heavily on bagasse in running the sugar industry a few years before the Head-Schemioth patent, in line with the global sugar industry at the time, in terms of fuel options. Mr Head must have been attributing to himself more than his fair share in the Egyptian biofuel innovation, inflating the impact of his engine by conjuring a more recent techno-energetic transition story. If we remember, the *Sugar Cane* had already reported in 1869 that the Rhoda sugar factory was running exclusively on bagasse, while coal fed other establishments. Head's patent then must have been eyed by Ismail as a potential improvement for his alternative fuel economy, not a radically new innovation. But it was indeed a great improvement, given the British MP's confirmation from the early 1880s that all agricultural production in Egypt was powered by biomass.[68]

As the engineers in Egypt ran tests for the cotton stalks in the Delta, they were simultaneously testing bagasse for Middle and Upper Egypt in parallel with the Head-Schemioth affair. At Cairo again, in March 1874, a test on a 10 hp portable engine yielded the result of 1 to 3 in consumption of coal to bagasse (by weight unit) for producing the same amount of horsepower.[69] During the discussion of his paper presented at the Royal Institution of Civil Engineers, Mr Head used this test result and two other tests from Egypt to show the modesty of his estimates on the economy of bagasse burning, a matter of great interest for the global sugar business and, as such, a great potential market for his engine. While presenting, he put forward numbers on the effectiveness of his invention: 2.5–3 lbs of cotton stalks, brushwood, or bagasse could produce the same amount of steam power as 1 lb of good coal. The ratio stood at 3.25–3.75 of wheat and barley straw to 1 lb of coal, and it was as low as 2 to 1 for dry peat to coal, and finally 2.25 to 2.3 for dry wood to coal.[70] To prove that he was not exaggerating, he mentioned the widely reported test conducted around

67 *Minutes of Proceedings*, Vol. 48, 86.

68 Stuart, *Egypt After*.

69 Lock et al., *Sugar*, 134.

70 *Minutes of Proceedings*, Vol. 48, 78; also copied later in the *Sugar Cane Magazine*, 1 January 1878, Vol. X, 31–8.

the same period by Mr Bramwell and Dr Letheby from the Khedival Supreme Sugar Commission. Under Khedival orders, they repeatedly tested bagasse burning on the *Daira Saniya* and came up with the average of 2.1 to 1 (bagasse to coal). Meanwhile, Mr Black, the chief engineer of the Maghagha sugar factory at Minya, Middle Egypt, and part of the *Daira*, came up with an even better result: 1.85 bagasse to 1 coal.[71] The disparity between Head's numbers (2.5–3 to 1) and those from Bramwell, Letheby, and Black was also transparently reported by Mr Head during that session at the Royal Institution. He cited the report from Bramwell and Letheby, which stated that if their tests were repeated using ordinary coal – not the good Welsh coal they used for comparison – and drier bagasse, the ratio would have gone above the 2.1 to 1 they reported.[72]

It did not stop at testing, as sugar refineries were already burning bagasse since before the Head-Schemioth patent. Mr Anderson from Easton & Anderson, who built the Aba Al-Wakf factory for Ismail Pasha in the early 1870s, was present at the discussion on Head's paper. He gave a long testimony, confessing that his initial design of the factory's tubular furnaces, ordered by his client to factor in the burning of bagasse exclusively, before the Schemioth-Head patent and Vienna Exhibition of 1873, was not efficient. The fire grate was too big. Mr Bramwell from the Khedival Sugar Commission, upon inspecting the delivered English furnaces, told Anderson that 'more than half of the grate was worse than useless'.[73] Technical modifications were ordered, coinciding with the production season of 1874, also the time when the Head-Schemioth invention made its big impact. On the modified method of feeding the bagasse into the middle of the fire, not from above, Anderson said that:

> the natives [stokers] squatted on the floor after their manner, with the megass piled around them, and pushed it down the mouthpieces with short sticks. *The fuel was put into the fire, as recommended by Mr. Head, not over it.* The effect of the alteration had been that, whereas before it had been necessary to use a considerable quantity of coal, during the season of 1875 virtually no coal was used. Coal had only to be burned until a sufficient quantity of megass could be collected; and

71 *Minutes of Proceedings*, Vol. 48, 99.
72 Ibid., 87.
73 Ibid., 97.

> at the end of the season a quantity of megass, quite equal to the coal consumed, was carried forward to the next season.[74]

Perhaps Mr Anderson did not report the full extent of the works for adapting the Aba Al-Wakf factory to exclusive bagasse fuel following the Head-Schemioth patent. We hear from elsewhere that a fully developed automatic feeding apparatus – fans, rollers, and all – was installed at the factory in the mid-1870s, adapted from the Ipswich stoking concept.

Such a mechanism was fitted to a tubular boiler at the Aba Al-Wakf factory. The dry bagasse was brought to the rolling tray that fed fuel to the furnace, that could be put to motion either manually or by steam. The same two fanning tools from the portable Head-Schemioth engine were installed at the mouthpiece of the furnace, which also rolled when the tray moved, bringing in the fuel. The same burning process ensued:

> By this means, a thin stream of fresh begass is continually injected into the furnace, so that the unconsumed fuel gradually forces the half-burnt and burnt begass towards the end where the combustion is most active. Another advantage is that, instead of throwing the fuel on to the top of the fire, which has been proved to be unsatisfactory when using vegetable refuse, the apparatus forces the begass first into the centre of the coldest part of the burning mass, thus supplying the fire with fuel, without checking the evolution of gases produced by the combustion in the middle and end of the furnace.[75]

Even after the British occupation of 1882, bagasse sustained the steam engines and boilers of the sugar factories that remained standing – sold as part of the bigger process of repaying the deposed Khedive's loans – and so these engines kept using bagasse as fuel, not coal. An American journalist who visited Middle Egypt in 1893 described a systematic process for this fuel economy, involving hundreds of peasants and the light railway system that spanned the sugar plantations.

> Railways run first to the drying-grounds for the megass – sometimes called the 'massash' – this being the hard fibre of the cane, which

74 Ibid., 98. Emphasis added.
75 Lock et al., *Sugar*, 133–4.

> having had the juice crushed out, is spread in the sun till it becomes combustible by the evaporation of its remaining moisture. A megass field is like a hay field, only vast in size, with white crushed cane spread out instead of hay, and hundreds of half-naked olive-skinned fellaheen as labourers. The engines and trucks rush about incessantly, for [. . .] there are sixteen separate steam-boilers adapted for burning fuel of large bulk and rapid combustion, and each [. . .] occupies three men night and day in feeding it.[76]

Burning More Coal, Always

So far, we have examined how the fuel economy of agrarian production – sugar refining, irrigation of sugar cane estates, and possibly comparable operations in the Delta – relied heavily on biofuel, roughly from the late 1860s to the 1890s. The question of coal, however, remains unaddressed. If Egypt deployed steam-powered machinery for agricultural production at a scale never witnessed before on the Nile in combination with the 1863 cattle murrain and the cotton boom in the wake of the American Civil War, what was the place of coal in this economy? Did coal burning slow down on the Nile? Can we learn a lesson on sustainability from this early Egyptian detour to other combustibles with – presumably – a lower carbon footprint? This is a question that demands longer-duration analysis.

Two points are necessary at the outset. First, almost all coal in Egypt across the nineteenth century and at least until the early twentieth century was imported from Britain.[77] In the mid-1870s, all coal imported to Boulaq harbour in Cairo came straight from Cardiff.[78] By 1916, coal

76 *Chambers's Journal* 10:479, 4 March 1893, 139.

77 In 1887–9 it was 99 per cent from England, and 1 per cent from elsewhere. See *The Statesman's Year-Book: Statistical and Historical Annual of the States of the World for the Year 1891*, London: Macmillan and Co., Limited, 1901, 1034. Earlier and later issues of *The Statesman's Year-Book* mention the same percentage.

78 The ships that brought coal returned full of cotton, sugar, wheat, or seeds. See McCoan, *Egypt*, 236; Alexander Delmar, *On the Resources, Productions and Social Conditions of Egypt*, The American Philosophical Society, [2 October] 1874, 11. Also, American coal was more expensive given the longer distance to Egypt: H. W. Mardon, *A Geography of Egypt and the Anglo-Egyptian Sudan*, London: Blackie & Son, Limited, 1902, 103.

imports from Britain were still forming the great majority of Egyptian imports, but American coal imports – the only other country exporting coal to Egypt by then – stood at around 13 per cent of British imports.[79] This means we can nearly be sure that almost all (if not all) of the coal in Egypt across the nineteenth century came from Britain, up until World War I.[80] For this reason, the numbers I bring forward shortly stop at the mid-1910s, the point after which British coal exports to Egypt ceased to be the only source of coal.

Second, since the opening of the Suez Canal in 1869, the share of the coal that went annually from British docks to Port Said to fuel the steamers that passed the canal formed a substantial chunk of the total imports we have on record. According to one estimate, in 1882, roughly 266,000 tons went for domestic consumption in Egypt, and 456,000 tons went for stoking steamers passing the Suez Canal with coal, at Port Said. The quantities were 360,000 to 726,000 tons in 1884, and it was 643,000 to 966,000 tons in 1896.[81] By the 1900s, Egypt was importing around 1 million tons of coal from Britain, and another 1 million tons went annually for stoking steamers passing through the Suez Canal.[82] This means that the numbers on record regarding the general British coal exports to Egypt in the nineteenth century can only show the general trend starting in 1869. Be it a ratio of one to two thirds or one to one later (burning of coal in Egypt versus stoking steamers), the analysis that follows shows general trends that need further exploration.

One final point to address before we move to the coal imports: when did the use of biofuel in agrarian production stop in Egypt? It was most likely phased out gradually by the turn of the twentieth century. In 1889, the British inspector of irrigation in Egypt, Sir William Willcocks, stated that for irrigation pumps, 'the fuel used is, ordinarily, coal, though, occasionally, one sees cotton and bean stalks. In Upper Egypt, where coal is dear, wheat straw is generally used.'[83] Elsewhere in his detailed report on the status of irrigation in Egypt after the British occupation

79 G. A. Roush, *The Mineral Industry, Its Statistics, Technology and Trade, During 1916*, London: Hill Publishing Co., Ltd, 1917, 142.

80 Charcoal was imported by Egypt from Turkey and Syria. See McCoan, *Egypt*, 168–9.

81 G. W. Steevens, *Egypt in 1898*, New York: Dodd, Mead & Company, 1898, 180.

82 Mardon, *A Geography*, 107.

83 William Willcocks, *Egyptian Irrigation*, London: E. & F. N. Spon, 1889, 238.

established control, he provided estimates on the weight and price of coal needed for the working of steam pumps and pumping stations, without comparable calculations on biofuels.[84] He also provided information on the use of coal for fuelling dredging machines at the Ibrahimiya canal.[85] This account from Sir William – which he compiled between 1884 and 1888 – suggests that: a) by the late 1880s, coal was the principle fuel for steam pumps and probably other machinery used in agrarian production; b) biofuels were still occasionally used to power steam pumps in the Delta, and much more frequently in Upper Egypt; c) there is no evidence from the British inspector of irrigation that the British administration had any designs on burning biofuels, which means that the practice probably survived at the hands of individual owners of steam engines, but it was no longer state policy. In 1903, the US Department of Agriculture conducted a survey of irrigation operations in Egypt, which found that 'coal is usually burned in [irrigation pumps'] engines, although cornstalks and straw are substituted for it in Upper Egypt'.[86] The reason was that coal's price increased the farther it was shipped away from the coast.[87] This might mean that the use of biofuels was almost completely stopped in the Delta by 1903 or earlier, then remained for some years in Upper Egypt.

However, since there is no reason to believe that coal prices decreased in Upper Egypt in later years, perhaps the use of biofuels remained a less frequent practice upstream until the arrival of the internal combustion engine and oil-fuelled agricultural machinery. There is, though, reason to believe that this latter development became mainstream by the mid-1920s. The monthly magazine *Misr Al-Haditha Al-Musawara* – which ran ads for automobiles and tyres starting in its first issue, published 25 October 1927 – ran a story in March 1928 on the benefits of 'modern ploughing machines' that were 'running on oil'.[88] A certain 'Tarzi Pasha' had been using an oil-run tractor on his estate in Upper Egypt since 1924, according to a letter written on his estate's letterhead that the Egyptian importers of McCormick-Deering

84 Ibid., 142, 240–2, 246.

85 Ibid., 216.

86 Clarence T. Johnston, *Egyptian Irrigation: A Study of Irrigation Methods and Administration in Egypt*, Washington DC: Government Printing Office, 1903, 44.

87 Ibid.

88 *Misr Al-Haditha Al-Musawara*, March 1928, 79.

tractors published as part of their advertisement.[89] Starting in August 1928, the same magazine ran advertisements for 'Petters Crude Oil Engines And Stone Irrigation Pumps', with capacities from 3 to 500 hp, all fuelled by oil.[90]

Moving on to the coal imports between 1850 and 1913, I will attempt to reconstruct the place of coal over the latter half of the nineteenth century, including during the years when biofuels were used (roughly late 1860s to 1880s).

Before the 1850s, records show that while Mehmed Ali kept searching for coal mines, he imported small quantities from Britain – negligible in some years – perhaps just enough to power the few Nile steamers and the handful of Cornish pumps he and his son Ibrahim brought from England by or slightly after 1845.[91] Abbas Pasha's years in power (1849–54) held the same trend (Table 9.1). Only by 1856–7 did the coal imports double, probably to power the Cairo–Alexandria railway (opened 1856), and perhaps also to fuel the steam-powered cotton gins that European capitalists started to establish in the Delta, roughly 1858–9 (see Chapter 4).

Table 9.1: British coal exports to Egypt and Egypt's share of the British global coal exports, 1851–77

Year	British coal exports to Egypt (tons)	British coal exports to Egypt (£)	British global coal exports (tons)	British global coal exports (£)	% of Egypt's share (in tons)	% of Egypt's share (in monetary value)
1851	62,560	24,714	3,386,430	1,302,473	1.8%	1.9%
1852	32,542	13,294	3,567,496	1,372,114	0.9%	1%
1853	18,447	8,726	3,935,062	1,504,591	0.46%	0.6%
1854	57,309	29,114	4,309,255	2,127,156	1.3%	1.4%
1855	41,538	21,027	4,976,902	2,446,341	0.8%	0.85%
1856	86,276	44,836	5,879,779	2,826,582	1.5%	1.6%
1857	115,091	59,380	6,737,718	3,210,661	1.7%	1.8%
1858	82,717	40,606	6,529,483	3,045,434	1.3%	1.3%
1859	111,290	54,305	7,006,949	3,270,013	1.6%	1.6%
1860	108,380	54,313	7,321,832	3,316,281	1.5%	1.6%
1861	118,492	58,458	7,855,115	3,604,790	1.5%	1.6%

89 Ibid., July 1928, 36.

90 Ibid., August 1928, 45. Same advertisement and more were published frequently in latter issues, until at least 1930.

91 Quantities for years 1842–5 respectively: 7,260; 13,000; 23,866; 48,063 tons. See Richard C. Taylor, *Statistics of Coal*, Philadelphia: J. W. Moore, 1848, 623.

1862	156,034	78,831	8,301,852	3,750,867	1.9%	2.1%
1863	264,958	123,138	8,275,212	3,713,798	3.2%	3.3%
1864	356,816	168,651	8,809,908	4,165,773	4%	4%
1865	388,828	206,091	9,178,477	4,427,177	4.2%	4.6%
1866	325,615	178,623	9,953,712	5,102,805	3.3%	3.5%
1867	417,601	233,143	10,424,888	5,400,353	4%	4.3%
1868	362,778	207,377	10,512,738	5,123,958	3.5%	4%
1869	495,043	262,956	10,245,350	4,830,521	4.8%	5.4%
1870	419,817	221,868	11,177,976	5,290,150	3.7%	4.2%
1871	480,915	248,542	12,208,009	5,879,680	4%	4.2%
1872	527,094	468,329	13,198,491	10,442,321	4%	4.5%
1873	547,549	600,862	12,077,507	12,370,638	4.5%	4.8%
1874	639,276	596,282	13,927,205	11,984,621	4.6%	5%
1875	533,492	402,809	14,544,916	9,658,088	3.7%	4.2%
1876	545,337	323,852	16,299,077	8,904,463	3.3%	3.6%
1877	520,476	282,268	14,880,899	7,477,699	3.5%	3.8%

Sources: *Statistical Abstract for the United Kingdom from 1856 to 1870*, fifteenth number, London: George E. Eyre and William Spottiswoode, for Her Majesty's Stationery Office, 1868; *Annual Statement of the Trade of the United Kingdom with Foreign Countries and British Possessions, for the Year 1855*, London: George E. Eyre and William Spottiswoode, for Her Majesty's Stationery Office, 1856; idem, *1860*; idem, *1864*; idem, *1866*; idem, *1869*; idem, *1871*; idem, *1873*; idem, *1877*. Quantities of global British coal exports in 1851–2 – since I could not locate them – were computed based on the average price of coal for the year 1853.

The upwards trend of coal imports held for 1860–2 as more steam-powered cotton-ginning factories were established, with the advent of the global cotton famine (Table 9.2; Figure 9.3). With the cattle murrain of 1863, and the mass importation of the locomobile – from England, for ploughing then irrigation in the Delta – Egypt's share of British global annual coal imports jumped from an average of 1.2 per cent for the years 1851–62 to 3.2 then 4 per cent (1863–4). The trend held until 1877, at an average of 3.9 per cent of annual global British coal exports going to Egypt.

Table 9.2: British coal exports to Egypt, 1851–1913

Year	British coal exports to Egypt (tons)	Year	British coal exports to Egypt (tons)	Year	British coal exports to Egypt (tons)
1851	62,560	1872	527,094	1893	1,432,998
1852	32,542	1873	547,549	1894	1,659,040
1853	18,447	1874	639,276	1895	1,539,754
1854	57,309	1875	533,492	1896	1,772,333

1855	41,538	1876	545,337	1897	1,580,289
1856	86,276	1877	520,476	1898	1,907,505
1857	115,091	1878	524,869	1899	2,125,924
1858	82,717	1879	596,484	1900	1,971,121
1859	111,290	1880	651,869	1901	2,093,424
1860	108,380	1881	876,945	1902	1,997,815
1861	118,492	1882	770,471	1903	2,174,620
1862	156,034	1883	968,982	1904	2,276,312
1863	264,958	1884	1,124,789	1905	2,274,368
1864	356,816	1885	1,142,516	1906	2,643,435
1865	388,828	1886	1,004,966	1907	2,981,468
1866	325,615	1887	1,268,961	1908	2,561,362
1867	417,601	1888	1,428,719	1909	2,651,400
1868	362,778	1889	1,465,654	1910	2,616,632
1869	495,043	1890	1,539,890	1911	3,160,735
1870	419,817	1891	1,578,036	1912	2,996,509
1871	480,915	1892	1,577,436	1913	3,263,936

Sources: *Annual Statement 1877* (1878); idem, *1886*; idem, *1891*; idem, *1896*; idem, *1900*; idem, *1903*; idem, *1907*; idem, *1909*; idem, *1913*. For the year 1879, only the exports of December were found, so I computed an approximate number for the year based on them; see *South Wales Daily News*, 29 January 1880. For the years 1880–81, *Cardiff Times*, 13 January 1883. For 1878, *South Wales Daily News*, 16 January 1879. For 1897, *Aberdeen Press and Journal*, 13 November 1897.

The two hikes in imported steam engines in the mid-1860s and around 1870 (Chapter 7, Figure 7.2) are matched in hikes in coal imports (Figure 9.3). The hike around the early 1870s in coal imports, however, peaked a few years after it peaked in the case of steam engines and machinery. While steam-powered machinery and engines reached a peak in 1871, the same happened with coal in the year 1874. The suspect here for this second hike is the sugar factories. The later peak in coal is probably because the factories took a year or two to be set up and start operating. Why would the sugar factories' operations become associated with more coal imports if they were powered by bagasse?

Figure 9.3: British coal exports to Egypt, 1851–1913

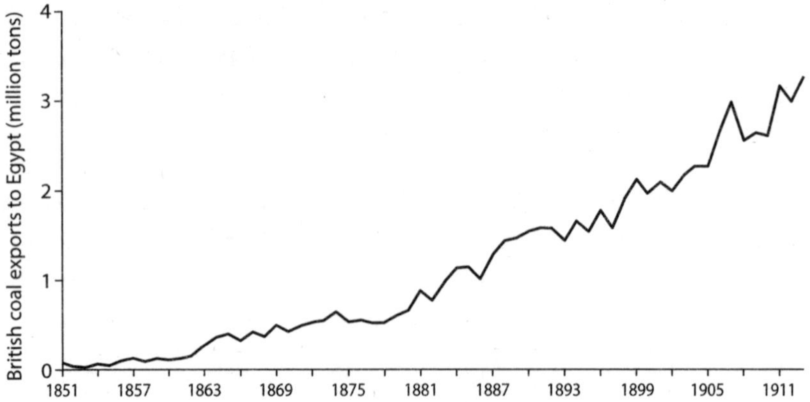

Source: Table 9.2

This is a crucial question. It is true that coal imports kept a steady pace slightly higher than 500,000 tons per year across the 1870s – after it peaked in 1874 – which might indicate that biofuel had a role in reining in coal imports for the new radically expanded sugar industry, but coal imports kept increasing anyway, even across the turbulent years of bankruptcy, starting in 1876, the Urabi Revolution, and the ensuing British occupation.

A probable answer here would be that, while agrarian production itself might have survived on biofuel, the transportation of the produced commodities (and perhaps raw material to the factories – for example, the light railway system in Middle Egypt built for sugar cane) kept running mostly on coal. This would include steamers carrying traders and commodities up and down the Nile, and the steamers stoked at Alexandria docks to take the commodities of sugar and cotton to their destinations in Britain, elsewhere in Europe, and across the eastern Mediterranean. The use of coal on board steamers and trains might have also been tied to the necessities of the mass and weight of fuels such as bagasse and straw of different kinds. These combustible materials – as we have seen – burned less efficiently than coal in terms of weight and volume versus the steam power they generated. Much larger heaps of bagasse or straw would have been needed to be transported manually to the trains and the steamers in order to run their trips. This would have required more labour power and time in comparison to labour power and time for stoking with coal. One would expect that a Nile steamer running on bagasse would make more stops for restocking – and this is just one potential logistical challenge. In the field, and at the factory, abundance of space and labour helped in softening this challenge.

In short, regardless of a transient shift to biofuel as an alternative to the more expensive coal – a production input that would have made the production of the commodities less profitable – the burning of coal carried on and kept increasing across the other links in the chain of trade in the commodities whose production was powered by biofuel.

Regardless of the details, the point is that the arrival of fossil-fuelled technology on the Nile – at an unprecedented scale since the mid-1860s – brought more fossil fuels to Egypt, in the long term, and during the time this fossil-fuelled technology was being deployed, even while it was being powered by alternative fuels. Fossil capital, expressed by the steadily rising imports of coal, kept expanding in Egypt.

Conclusion: How It Burned

From Ukraine to London and Ipswich, from burning straw in 10 hp small portable steam cultivators to running the boilers and engines of enormous sugar factories in Middle and Upper Egypt, a nineteenth-century transition was underway, to biofuels, for agrarian cash crops and agro-industrial production. This process involved substantial Egyptian participation in the global knowledge networks of steam technology, where the Khedival Supreme Sugar Commission tested cane trash, cotton stalks, and other combustible non-fossilised material for burning capacity in locomobiles and sugar-refining boilers. The process involved the tweaking of the boilers of the cutting-edge sugar factory of Aba Al-Wakf, imported from Easton and Anderson of London, and by the turn of the 1880s, several witnesses assure us, the fuel economy for producing sugar and other crops in Egypt did not need coal, even while Egyptian coal imports kept rising steadily. Head's vision for globalising steam power outside sites rich in coal endowments was fulfilled: 'The only method of rendering the use of steam power universal, particularly for agriculture, would be to construct the boiler of the engine so as to utilise the local supplies of *combustible material of every kind*.'[92]

92 Head, 'The Combustion', 60. Emphasis added.

With the detour this chapter took into the fireboxes of steam engines of the steam cultivators and sugar-refining boilers, we stand at odds, seemingly, with Andreas Malm's account of how fossil capital ignited the fire of the Capitalocene from England. In his account, coal was abundant underneath English soil. The steam engine was invented to bring more of it to the surface of the earth. Then the steam engine found its way, fed by coal (always), to the factory, to replace water power, as it proved more aligned with the imperatives of capital accumulation.[93] When Malm offered a theory on globalised fossil capital, he looked at China, when it used its rich coal endowments to spur industrialisation at the turn of the twenty-first century.[94]

This account of how the fossil-fuelled technology made its way to 1860s–70s Egypt, however, unfolds as a story of igniting steam engines to power expanded agrarian commodity production – cash crops – in the absence of coal, which was prohibitively expensive for economic calculations on producing the commodities the steam engines were imported to produce. This was not much unlike the situation in Ukraine and the Russian Empire, and not different, perhaps, from the situation in many other sites bent on growing more cash crops at the time using steam power. One is tempted then to assume that the availability of alternative fuels, harvested at the surface of the earth, and fed to the fireboxes of locomobiles through Head-Schemioth implements or other comparable feeding mechanisms that preceded it by some years in Egypt, had a role in encouraging the use of even more steam engines, not fewer. Fossil capital – the latching of engines originally designed to burn coal to the means of production, and the expansion in profits through the accumulation of carbon in the atmosphere – was not necessary from this perspective for the rise of industrialisation proper known to be intimately associated with the steam engine.

But, then, the formation of a hypothesis to that end – that is, to dismiss fossil capital as an irrelevant concept for the socioecological history of modern production technology – seems to have at least two flaws. I offer two notes towards addressing them.

93 Andreas Malm, *Fossil.*

94 Ibid., 290.

First, the soil is non-renewable, which means that those early attempts at burning biofuels instead of fossil fuels were not going to last long.[95] As the debate on bagasse as fuel versus fertiliser from the *Sugar Cane Magazine* shows, using crushed sugar cane shoots as fuel, instead of returning them to the soil as fertiliser, had a negative impact on the soil under sugar cane, eventually producing seemingly healthy canes with less sugar content. The same process can be gleaned from Egypt with a layer of complexity added. Ismail Pasha, for example, ordered the building of pigeon towers across Middle and Upper Egypt once he turned to using bagasse as fuel, to make use of the pigeon guano to manure the sugar cane fields, a manure that replaced bagasse-as-manure. Half a ton of this guano was used per acre. It might have been a systematic process in Middle Egypt, where and when 'the keeping of a flock of pigeons [was] part of the fellah's duties to the state'.[96] The pigeons fed 'one way or another, from the fellah's corn-field'.[97] By 1879, a famine raged across Upper Egypt, taking the lives of no fewer than 10,000 people. The failed Nile inundation of 1878 and severe taxation were blamed.[98] However, the process of cultivating sugar cane across Middle and Upper Egypt over that decade to feed Ismail's twenty modern sugar factories must have had a share too in the then-growing scarcity of grains across the region, and perhaps the pigeon towers for manuring the sugar cane fields shared in some responsibility too, as the pigeons – in huge numbers to produce the half-ton of manure per acre of sugar cane – also ate grain. In other words, dedicating almost all of the organic material the Khedive could put his hands on – that originally manured the land – to feeding steam engines for producing cash crops and refined white sugar could have understandably released complicated ecological processes that reduced the fertility of the land and starved the people feeding off it. The story of the pigeon towers is just one illustrative example of the chains of causality that the use of alternative fuels could have unleashed.

95 It takes hundreds to thousands of years for one centimetre of soil to form. See Hanna Holleman, *Dust Bowls of Empire: Imperialism, Environmental Politics, and the Injustice of 'Green' Capitalism*, New Haven and London: Yale University Press, 2018, 9.

96 Delmar, *On the Resources*, 12.

97 Ibid.

98 *The Times*, 30 March 1878, cited in 'The Famine in Upper Egypt', *Ayr Advertiser*, 17 April 1878.

In the absence of coal, the subterranean fossilised plant matter powering steam engines while leaving the surface of the agrarian land itself intact – only insofar as coal combustion meant leaving the plant matter to return to the soil as manure – we can assume that Khedival Egypt's burning of above-the-ground plant matter for powering steam engines was extremely unsustainable to the land, even more than coal, with short-range consequences, that is, shorter in range than those of coal combustion causing global warming, or coal causing the deterioration of soil by means of more intensive cultivation practices, and so on. It would not have been possible to sustain the practice for long, given the limited availability of land hinging on proximity to the Nile – in other words, it was not possible to clear more woods or forest; there were simply none there. It must, then, be seen as a transient fix to the problem of the absence of coal outside coal-rich agrarian production sites of the world that wanted to use steam engines. Perhaps this is part of the reason the Head-Schemioth patent is nearly absent from the literature on the Industrial Revolution: it did not make a big enough dent to be noticed, as it quickly declined and was forgotten.

Second, the logic of capital accumulation within which fossil capital operates is more than part of the equation for the valorisation of value through the burning of fossil fuels, where the latter get fused to the capitalist's means of production. Fossil capital is also a process: 'an endless flow of successive valorisations of value, at every stage claiming a larger body of fossil energy to burn'.[99] In 1860s–70s Egypt, generally fossil-fuelled technology was operating without fossil fuels to power the production of sugar and possibly other commodities, like cotton. Describing this technology as newly acquired fossil capital – in the absence of fossil fuels – is a contradiction of terms. However, as the coal imports from Britain show, fossil fuels in Egypt kept rising during the nineteenth century and into the twentieth, even if the production of some of the leading commodities produced through this agrarian mechanisation stayed mostly coal-free for a few decades. It was inevitable, once more quantities of white sugar and cotton were produced, that more fossil fuels would be used. In the lands of sugar, for example Middle and Upper Egypt, with each new sugar factory built, miles upon miles of light railway tracks were built. The modern sugar production

99 Malm, *Fossil*, 290.

line was at the heart of transportation networks powered by coal. The sugarworks complex Ismail established in the 1870s was known in the industry as *usine centrale*, the central factory, a naming that entailed that its manufacturers also specialised in building railways for transporting production inputs to the factory, like sugar cane, straw, and bagasse from the fields. Nile steamers for transporting production inputs to the factories and for taking the sugar to the coast downstream were essential too. Coal for the steamers taking sugar and cotton to Europe and elsewhere, and bringing in coal to power more steamers and trains across Egypt, could have accounted for a substantial part of the increasing quantities of coal imports. In other words, the diffusion of biofuel-based steam power opened the gates to dependency on coal-based steam power in Egypt, leading to more coal burning. Ultimately it was not a sustained transition to biofuel-based steam power. It was biofuels deployed just as further fuel alongside the increasing fossil fuels, as the coal imports statistics hold witness to.

Another even more essential aspect of this process is that it was always a global process, not confined to agrarian Egypt of the late nineteenth century. All analysis of the global process of the Capitalocene – predicated in the case of Egypt on the spread of fossil capital, in combination with loans, leading to colonialism – must be of a global nature. The flows that the Egyptian agrarian commodities of cotton and sugar were part of included the factories of England burning coal in steam engines to produce textiles (from the cotton) and an important cheap asset for the reproduction of the British labourer (sugar).[100] The financial capital used to buy the locomobiles and the sugar factories was accumulated through the burning of this coal in England and elsewhere, and this financial capital was then lent out to Ismail Pasha, incurring interest that was put to use in expanded capitalist endeavours in Europe and elsewhere. From the perspective of Egypt, it was an intensification in using all sorts of energy at hand in general during this phase of what we can call the proliferation of proto-fossil-capital that walked hand in hand with the horizontal deployment of the infrastructure necessary for

100 This is a brutally simplified example. We are not sure that British labourers consumed Khedival sugar, but in the larger global scheme of sugar production, this sugar must have contributed to making all white sugar everywhere more affordable. The same goes for the cotton and English textile mills.

industrial capitalism to take hold. The factory, the railway, and, later, the organisation of labour in the factories were some of the manifestations of this phase of deploying proto-fossil-capital, or fossil capital making its way to a coal-free part of the Global South.

Fossil capital, in Malm's definition, as a global process for endlessly valorising value and claiming over time a larger body of fossil energy to burn, incorporated Egypt into its ever-expanding circle, even though its production of sugar and cotton was sustained for a while by means of bio-fuelled steam power.[101] Perhaps the same can be said about the production processes in which the Head-Schemioth patent was used across the globe in the late nineteenth century.

This is how it burned, the steam-powered engine, as it started to take hold in precolonial Khedival Egypt. It burned through adding cane trash and cotton stalks and the straw of grains to the fire to ignite the newly imported machine used in agrarian mass-commodity production. It burned through the huts and houses of the Qau rebels, whose villages were razed by Ismail's army while they lay by the dozen waiting for the chopping axe to fall on their necks, because they challenged their exploitation in coerced labour for digging canals to install more steam irrigation pumps and grow more cash crops. It burned through elaborate ploys over political power that was the target of the Khedival monopoly over the steam engine in the early 1860s, and the confiscation of lands from the peasants as much as from prominent Viceroyal family members, like Prince Halim and Mustapha Fadil Pasha, Ismail's younger brother. It burned through the millions of pounds taken from Europe as loans, which eventually brought the British navy to Egypt, where they stayed for decades. Combustible materials of all kinds, combustible carbons, actual and metaphoric, plant matter, coal shipments reaching Egyptian shores more each year than the earlier, bank notes, cannon fire of the British fleet, and bodies of millions of peasants, burned as steam power extended over the Nile valley.

101 Malm, *Fossil*, 290.

PART III

A Fable: Capitalism and Its '*Afarit*

10

'Don't you see, *Pashmuhandis*, that we are drowning?'

' الحمدُ لله الذي مَعادِن خَزائنه لا تُحصَى' .

Praise be to Allah, whose minerals – in His safes – are beyond count.
Rifa'a Al-Tahtawi, in the introduction of a volume on minerals translated from the French (*Minerals Useful for the Livelihood of the Population*), before a French–Arabic dictionary existed in Egypt

This is how it ignited, and how it burned: steam power on the Nile. But how did the people make sense of this power? This is a broad question that requires a different set of archival materials, and one has to recognise that the viewpoints of the subaltern on steam power can hardly be retrieved at all. But, since we are after the history of global warming, the Capitalocene (fossil-fuelled technologies that powered capitalism and catastrophically changed the state of the planet), I can limit my inquiry here to the views of those who accounted for the economic power of the fossil-fuelled technologies and those who felt its heat. So, in this chapter I briefly turn to pointers on the cultural history of steam power's arrival in Egypt, from the viewpoint of elite intellectuals associated with the ruling class and the peasants. As it happens, the view of both groups fleshed out a Capitalocene in the making: capitalism and the dangers coming with its means of industrial production. Between Al-Tahtawi's opening of perhaps the first book translated to Arabic in Mehmed Ali's Egypt ('Praise be to Allah, whose minerals – in His safes – are beyond count') and the peasants' frightened plea to an engineer, once he ignited one of the first locomobiles in the Delta ('Don't you see, *Pashmuhandis*, that we are drowning?'), to shut it down, this chapter elucidates what it meant for Egyptians in different positions to turn on the power of steam.

'Afarit, Industrialisation, Colonialism, and Global Trade

The tension between the positions of the elite and the subaltern can be made clearer if we turn to engineer, educator, and minister of public works Ali Mubarak. In the year 1882, he published the first volume of four, of his pedagogical text, *'Alamuddin*. The title is the name of a fictional Azhar sheikh who became friends with an English orientalist. Mubarak used the setup of these two worlds meeting to write lessons about science and culture. The first lesson in volume I of *'Alamuddin* – targeting the growing population of school students – is on steam technology. It is narrated on board a train taking both men, and *'Alamuddin*'s son, to Alexandria, from where they were to board a steamer to England. On the train, the Azhar scholar told the English gentleman that some laypeople say that the train was powered by *jinn* and *'afarit*, creatures of the underworld.

The English gentleman starts a long lecture on steam technology, as the train moves fast across the Delta. It begins with the concepts of atmosphere, atmospheric pressure, and bar, then moves on to the measuring units of the metric system, compared and measured to Egyptian units of length and weight. Next, crucial to dispelling belief in otherworldly motive powers in the boilers and fireboxes, come the basics of the steam engine's thermodynamics, which had no Egyptian counterparts. The English gentleman then narrates the historical development of the steam engine, before he changes topic to trains, their economic and social benefits, and relative safety.[1]

The first scientific lesson of this pioneering encyclopaedia of popular science and arts, written by a prominent state official, educator, and one of the very first Egyptian chief engineers in the nineteenth century, was

1 Ali Mubarak, *'Alamuddin, Vol. I*, Alexandria: Almahrousa Press, 1882, Chapter 7 ['*Al-musamara al-sabi'a*']. For recent deliberations fleshing out other perspectives on *'Alamuddin*, see On Barak, *On Time: Technology and Temporality in Modern Egypt*, Berkeley: University of California Press, 2013.

the story of the steam engine and thermodynamic metrics, to expel the superstitions of the locals who suspected that the underworld was operating there instead of physics, as much as because thermodynamics and the steam engine were perhaps the most useful scientific topics of the age, and as such came naturally before other topics.

Rifa'a Al-Tahtawi, too, the first translator in modern Egypt who translated a volume on mining before a modern French–Arabic dictionary existed, was fascinated by the steam engine.[2] Like Ali Mubarak, who explained to the school students and learned Egyptians the concepts of steam power and thermodynamics, he wrote a poem about the machine powered by *wapour*, or *vapeur*, from French. He described a kind of ideal-typical 'steamer' that was to be found in the trains and steamboats as 'mind-blowing', 'the sun's yellowness fades beside its heat, it shies away when the *wapour* blows its whistle', 'its steam smells of amber', 'a volcano of fire when it rumbles', 'disdains the wind when it blows, so much slower than the speed of the *wapour*', and 'it gave Ismail global prominence [when he brought it]'.[3] Al-Sayed Salih Magdy, one of the earliest translation students of Al-Tahtawi, who learned French at the school the latter established in the 1830s, accorded Sa'ed Pasha (1854–63) a eulogy, in allegorical comparison with a Nile steamer, where both shined with power, beauty, and magnificence.[4]

Al-Tahtawi was among the first few dozen Egyptian students Mehmed Ali sent to learn in Paris in 1826. By the 1860s – after he established one of the first higher education institutions in the 1830s – he was a very influential figure in shaping Egypt's educational policies, and a big landowner. While Mehmed Ali is described as the 'Builder of Modern Egypt', Al-Tahtawi is dubbed as the 'Pioneer of Egyptian Enlightenment'. The book that featured his poem on the steam engine, *Manahij Al-Albab*

2 If there were earlier French dictionaries in Egypt at the time or earlier, Al-Tahtawi was not aware of one. On earlier translation endeavours from European languages, starting in the 1810s, see Peter Hill, 'The First Arabic Translations of Enlightenment Literature: The Damietta Circle of the 1800s and 1810', *Intellectual History Review* 25:2, 2015, 209–33.

3 Rifa'a Al-Tahtawi, *Manahij Al-Albab Al-Masriya Fi Mabahij Al-Adab Al-'Asriya*, in *Al-A'mal Al-Kamila Li Rifa'a Rafi' Al-Tahtawi* [Complete Works of Rifa'a Rafi' Al-Tahtawi, Vol. I], edited by Mohamed Emara, Cairo: Maktabat Al-Usra, [1869] 2010, 109–10 (lines 1, 6, 14, 16, 25, and 30).

4 Al-Sayed Salih Magdy, *Diwan Al-marhoum Al-Sayed Salih Magdy Bey [The Complete Poems of the Late Al-Sayed Salih Magdy Bey]*, Cairo: Boulaq Press, 1893, 1–2.

Al-Masriya fi Mabahij Al-Adab Al-Asriya (roughly: Methodologies for the Egyptian Mind for Navigating Contemporary Thought), introduced an unprecedented body of theoretical deliberations on industrialisation, commodity production, free trade by colonisation, and global trade, among many other political economic themes.

While Mubarak addressed the general public about the hard science of the engine, Al-Tahtawi lectured them on the useful economic processes and structures the engines were enmeshed into. The central concept that the book advances is *al-manafi' al-'umumiyya*: the common good, or public wealth, which one can easily stretch to 'the wealth of the nation'.[5] The wealth of a nation, according to Al-Tahtawi, comes from three sources: 'agricultural, manufactural, and commercial', in an interconnected sense, one leading to the other.[6] He adds that in France (where he had his education, 1826–31) the wealth of a nation is called *industrie* (which he transliterated to Arabic characters, an act that shows how 'industry' was for him much broader than the technical processes of manufacturing).[7] *Industrie* (or *industriyya*, the French transliterated-to-Arabic word) shines in different shades of 'industrialisation' as he understood it, and as industrialisation took its place in various niches across the nineteenth century. It meant, in the literal sense, 'achieving higher levels in dexterity and skilfulness' or industriousness, and turning 'primal material [of agrarian products into] something that can be utilised [i.e. industrial processing]' and sold on the market – producing

5 The lineages of Al-Tahtawi's economic thinking are not discussed, but he said in *Manahij*'s introduction that the book mixes together European and Arab sources of wisdom and information. Ghislaine Alleaume convincingly shows that *Manahij* has much in common with the syllabus of a certain 'industrial economics' course put together and taught by the Saint-Simonian engineer Charles Lambert at the Egyptian engineering school (*Muhandiskhana*) in the 1840s–50s. If Al-Tahtawi factored in Adam Smith in this book, it must have come filtered through Saint-Simonian writings. It is noteworthy that Ali Mubarak inherited the management of *Muhandiskhana* from Lambert, who left in the early 1850s. See Ghislaine Alleaume, 'L'Économie politique saint-simonienne et les élites techniques de l'Égypte moderne', in *Études saint-simoniennes*, edited by Philippe Régnier, Lyon: Presses universitaires de Lyon, 2019. Al-Tahtawi/Saint-Simonian connection on *industrie* is explored further in Amr Khairy, 'Industrie and Al-Industriyya: Translating "Industrialization" in 1860s Egypt', *Capitalism: A Journal of History and Economics* 6:1, 2025.

6 Al-Tahtawi, *Manahij*, 327–8.

7 Ibid., 433.

the commodity, in other words.[8] After framing it as an economic whole embodied in the commodity, *industrie* for Al-Tahtawi is, next, a process that unfolds through colonialism, to open markets for the land-based commodity. He says, tracing colonialism in the early modern world, that 'the Spaniards, the Portuguese, the French, the Flemish' did commit the progressive act of 'colonial *industrie*, or [i.e. which also means] international trade'.[9] He then commends Britain for 'opening up expansive lands across vast distances', including in India, to bring 'plentiful benefits to the people of that Kingdom [Britain] and to the people of other kingdoms by extension'. The circle of steam-powered industrialisation becomes complete when nations, like Britain, 'labour towards increasing [their] trade'.[10]

It Burns, It Explodes, It Flies, It Eats the Young

The laypeople who rationalised the steam engine as the dwelling place of *'afarit*, creatures of the underworld, had good reasons for invoking this otherworldly power to make sense of a source of danger that occasionally tore the fabric of their material life in absurd and frightening scenes.

Max Eyth, the chief engineer of Halim Pasha and the agent of the Fowlers, the Leeds engine manufacturers, was setting up his second locomobile in the Delta at some point in 1863 when such an incident occurred. It is a rare eyewitness account that opens a window into exactly why the peasants thought *'afarit* were controlling the steam engine. It was a 10–20 hp steam pump bought and set up to irrigate 1,000 acres in Prince Halim's estates in Thalia, Munufiya, in the middle of the Delta. Despite the imposing mass of the locomobile, which the peasants were perhaps seeing for the first time, the air was filled with joy and anticipation. Children jumped to swim in the canal close by, women rushed to the English machine once it was ignited and started pumping water, ululating joyously on the way. A hundred men stood with wide grins on their faces. Perhaps they saw in the tremendous power of the

8 Ibid.
9 'Al-Industriyya al-Quluniyya' in Arabic. Ibid., 437.
10 Ibid., 437, 451.

machine a benign 'miracle'.[11] The mayor of Thalia, Ali el-Naggar, was sitting beside the engine, drinking his second cup of coffee as he witnessed the miracle, when the engine operator, a French mechanic, decided to push the machine to go faster, perhaps to demonstrate its might. The canal the engine was irrigating quickly filled to the brim with water. It overflowed, and in a few minutes over ten acres of cotton were flooded. The peasants rushed with their baskets, filled them with dirt to dam up the canal, and the mayor Ali el-Naggar pleaded to Eyth, 'Stop, by the All-Merciful, stop! Do you not see, *Pashmuhandis* [i.e. chief engineer], that we are drowning? The dams out there are broken, the earth is floating away, the field is under water! Ten feddan [acres] of cotton are perishing. Stop, *Pashmuhandis*. A devil is in your pump and will drown us all.'[12]

The *'afarit*, and their overlord the devil, were invoked for other even more compelling reasons for the peasants. By the 1890s, we get reports on engine accidents of spectacularly tragic descriptions. In October 1894, in Qwisna, Munufiya, the boiler of a steam-powered grain mill, beside the train station, exploded. It tore into two pieces, and one of the two flew 200 metres in the air to land on the other bank of the Nile and the explosion destroyed several neighbouring buildings. 'The building crumbled and around 30 persons died', including twelve persons who died instantly from the explosion, most of them women.[13] In October 1891, 'a huge fire erupted because of the engine in Housh Sharqawi in Cairo. Another engine in Batniya, Cairo, exploded, and the fire spread across several blocks. In June, 'a vicious accident' occurred in the peasant community of Shubak, Giza, where 'some of those killed were hit by projectiles that flew over 150 metres from the centre of the engine's explosion'. In Alexandria, in early 1892, an engine's boiler at Ibrahimiya exploded and flew, across the skyline of a residential area, for hundreds of metres then penetrated through the walls or ceiling of the house of Mohamed Al-Sawah, who died, along with many others.[14]

11 Max Eyth, *Der Kampf um die Cheops-Pyramide* [The Battle for the Cheops Pyramid], Heidelberg: Carl Winter's Universitätsbuchhandlung, 1902, Chapter 1.

12 Ibid.

13 *Al-Mu'ayyad*, 14 October 1894, 3. *Lisan al-Hal*, 22 October 1894, 1.

14 All accidents listed in *al-Mu'ayyad*, 10 October 1892, 2; *al-Mu'ayyad*, 16 July 1892, 1.

Aside from large accidents of explosion and fire, there were the sneaky, silent kind of accidents. A woman's hair was caught in a grain mill's belt in July 1906. Her body was caught in the wheels and she died. The technician who heard her screaming and tried to save her lost his arm.[15] The same kind of accident was still being reported frequently in the spinning mills of Mahalla Kubra in the 1920s, now that oil replaced steam, but the belts kept taking the lives of peasants who were hired as unskilled workers.[16] In 1893, another engine accident unfolded like a nightmare. At a major steam-powered sugar factory that hired thousands of peasants, somewhere in Middle Egypt, a technician lost his grip on a wrench. It went down a trap door into an engine room under the ground. Unaware of the poisonous concentration of carbon monoxide in the poorly ventilated chamber, he sent a young labourer to fetch the tool. The boy fell and died but the technician did not realise it, so he sent another boy, who also died. Then he went in himself and died. It all happened in a few minutes. The wives and mothers of those who might have been some of the first martyrs to industrial carbon particles in Africa came in the evening to collect their bodies. They attacked the factory, throwing stones at its facades and walls of iron and glass.[17]

In the 1920s, the spinning mill peasants-turned-workers in Mahalla Kubra used to say matter-of-factly when they lost their comrades in machine accidents that the engines needed the blood of boys, human sacrifices for the machines to operate.[18] A short story by Yahia Al-Tahir Abdallah, the most prominent modern novelist and short-story writer from Upper Egypt, depicted rural populations' fear and objections to a mechanised flour mill coming to a village close to Luxor in the 1940s, because, 'for as long as we lived, we never saw a machine that ran without first eating a young boy'.[19] Up until the present, it is a tradition some observe to slay and distribute the meat of a cow or a sheep on buying a new vehicle, and the slaying of the animal usually happens right beside

15 *Al-Ahram*, cited in *Lisan al-Hal*, 25 July 1906, 1.

16 Fikri Al-Khouli, *Al-Rihla* [The Journey], Cairo: Dar Al-Karma, [1987] 2018, 14.

17 *Chambers's Journal 10*, 4 March 1893, 138–9.

18 Al-Khouli, *Al-Rihla*, 14.

19 Yahia Al-Tahir Abdallah, *Tahouna Al-Sheikh Mousa* [Sheikh Mousa's Mill], short story in *Al-A'mal Al-Kamila* [Complete Works], Cairo: Dar Al-Mustaqbal Al-Arabi, [1970] 1994, 38.

the vehicle. Perhaps this is connected to replacing the human sacrifice for the engine with an animal sacrifice – appeasing the machine's appetite for the blood and flesh of the young, the future generations.

'A devil is in your pump and will drown us all'

The connection between the coming of the steam engine through explosions and fire, through stealthy death by carbon monoxide and moving belts, and its rationalisation in rural Egypt as demonic makes more sense now. These incidents and tales can be read as interpretations of the dark side of steam power that Ali Mubarak presented as a thermodynamic scientific process in his encyclopaedia 'Alamuddin. They can be particularly seen in their tensions with – and a dark side of – industrialisation and a certain capitalism that expands by colonialism, as explained and advocated by Rifa'a al-Tahtawi as the sum of the wealth of the nation. He explained the new age of the Industrial Revolution – while poeticising his fascination with the tremendous power of steam – as the age of gearing agriculture, manufacturing, and commerce to produce and sell commodities (*industrie* in a word) including by means of colonising new 'unindustrious' lands ('colonial industrialisation or international trade', in his words). These are two faces of fossil-fuelled technology hooked to capitalism: unprecedented levels of producing commodities by subsuming labour and nature, and plundering both, leading to demonic, disastrous outcomes.

But there is more to this than seeing peasants and laypeople 'demonising' the engine because it came with unexplained kinds of danger; they could in fact have got it right, as the anthropologists of magic may tell us. Magic exists in modernity, and as with premodern beliefs it has great interpretative value for understanding how people make sense of historical and contemporary change.[20] Peter Geschiere argues for

20 Peter Pels, 'Introduction: Magic and Modernity', in *Magic and Modernity: Interfaces of Revelations and Concealment*, edited by Birgit Meyer and Peter Pels, Stanford: Stanford University Press, 2003, 3. For invoking magic in problematising modern science and technology, see Laura Nader, 'Epilogue: The Three-Cornered Constellation', in *Naked Science: Anthropological Inquiry into Boundaries, Power, and Knowledge*, edited by Laura Nader, London and New York: Routledge, 1996; magic in reimagining and reinterpreting technology and the global economy: Alf Hornborg,

reimagining magical beliefs: instead of seeing in them static primitive belief systems, we can look to them to tell us how 'modern changes have been grafted on local realities'.[21] Michael Taussig goes even farther: 'Magical beliefs are revelatory and fascinating not because they are ill-conceived instruments of utility, but because they are poetic echoes of the cadences that guide the innermost course of the world.'[22]

If the tales in this chapter are re-read – beyond the interpretive approach for understanding where they came from and why they originated – rather like fables about fossil fuels and the machines they powered, what are the lessons learned regarding our contemporary predicament of global warming? How can we read them as 'poetic echoes of the cadences that guide the innermost course of the world'? A fable, after all, means, among other things, 'a narration intended to enforce a useful truth'.[23] Here, I stop with three notes towards collecting such lessons.

The first lesson is to be found in the paranormal protagonists of the tales, the *'afarit* and their natural habitat, the underground. Fossil fuels also emit great amounts of heat energy when they burn to power things, to move and act, and they come from the underground. In the Muslim tradition *Al-Shaytan*, the devil, is the supreme leader of all *jinn*, including *'afarit*. He was originally an angel created of fire. When Allah created Adam and asked the angels to bow to him, they all abided, but the devil refused, saying he will never bow to a creature of mud, a material inferior to his flesh of fire. Allah banished him from heaven for eternity, and his sin was arrogance. He told Allah to wait and see what he would do to humans until Judgement Day. Until then, he promised his creator, he would keep intriguing and leading humans astray. He then took *jinn*

Global Magic: Technologies of Appropriation from Ancient Rome to Wall Street, New York: Palgrave Macmillan, 2016. Jean and John Comaroff, 'Occult Economies and the Violence of Abstraction: Notes from the South African Postcolony', *American Ethnologist* 26:2, 1999, 279–303 (on magical beliefs in the context of South Africa's neoliberal turn); reinterpreting magic in the diffusion of capitalism to nineteenth-century West Africa: Peter Geschiere, 'Sorcellerie et Modernité: Retour sur une Etrange Complicité', *Politique Africaine* 79:3, 2000, 17–32.

21 Peter Geschiere, *Witchcraft, Intimacy, and Trust: Africa in Comparison*, Chicago: University of Chicago Press, 2013, xiii, xxviii.

22 Michael Taussig, *The Devil and Commodity Fetishism in South America*, Chapel Hill, NC: University of NC Press, [1980] 2010, 15.

23 Merriam-Webster dictionary online.

and *'afarit* as aides. Looking at fossil fuels, we see these things from the underground burning steadily away, and we are oblivious to the climate tipping points fast approaching – with capitalism, and fossil capital, leading humanity astray. The human agents and the specific anthropogenic processes particularly benefiting from burning fossil fuels and emitting carbon become the *'afarit* leading humanity astray to its extinction, moving rapidly to a secular Judgement Day at 3 or 4 degrees, or higher, above pre-industrial levels. Rather like a Quranic or Faustian pact with the devil: regardless of the material progress and benefits delivered, always to some only, the price is terrible.

The subject of this second note is also terrible. The dark stories of sacrifices of the children to the fossil-fuelled engines in Egypt on one hand, and the temporal trajectory of global warming on the other: the unsustainability of the business-as-usual that sacrifices future generations. In the cosmological arrangement of turn-of-the-century Egyptian peasants, either the younger generations were imagined as being sacrificed to the engines, or there was fear about them falling prey to the same fate. The engines were said to be hungry for the lives of the young ones. The fireboxes and combustion spaces of the machines were the haunts of *'afarit* causing or metaphorically replicating the perpetual march of the new fossil-powered technology and its associated transitions. In our moment of global warming, we have something eerily similar happening: children, the young, those born and those still unborn, are victims of the fossil-powered engines and fossil capital more broadly. The longer fossil fuels are allowed to power engines, and the longer the Faustian deal holds, the more children of the future will fall victim to the consequences of these processes. As long as fossil capital feeding capitalist production keeps expanding – the Capitalocene in a word – those knowingly running this establishment are sacrificing children to the engines.

The final analogy I highlight addresses what we can do about this chilling image of the premeditated child sacrifices committed by the actors perpetuating global warming. Here, the peasants of Thalia and their mayor, Ali el-Naggar, are my source of inspiration. When the mechanic got carried away and fed the steam pump with ridiculous amounts of coal to demonstrate the power of his machine, the result was the rise of water levels from the canal and the drowning of ten acres of cotton within minutes. The peasants and their mayor had the obvious reaction: they rushed with their baskets to wall up the canal to mitigate the danger, and the mayor demanded that the source of danger be shut down: switch off the devil-haunted engine of the water pump, or more land will be flooded! Some comparable obvious facts about planetary risks of global warming were stated recently in a climate science meta-analysis. Kemp et al., in their call for a research agenda more responsive to bad-to-worst-case scenarios of climate change – the underestimated endgame scenarios – stated that a risk is more than the hazard itself; it also includes 'vulnerabilities, and exposure, but also responses'.[24]

The people of Thalia were responding to the hazard they saw with their own eyes, and their leader – who, arguably, was the only native person with the power to override the decisions of the machine operator – demanded that the fossil-powered machine be stopped. In our global warming, however, our view in terms of the risks is distorted. The impact of the fossil fuels and their machines is denied, and the mayors of the world, in a scene that belongs to the theatre of the absurd, just keep sipping their cups of coffee, negotiating net-zero, carbon trade, and futile – rather magical – carbon capture schemes at COPs, while the machine operators keep flooding more and more acres of the finite planet. The 'epistemic risks are serious concerns for public health crises',[25] as Kemp et al. state, reminding us of the obvious. Global warming, the source of the greatest morbidity and mortality crisis facing humanity – and eventually climate system breakdowns – remains shrouded in epistemic shawls of denial and inaction.

We might want to hear Ali el-Naggar one last time before we turn the page of the 'lessons learned' from the *'afarit*:

> Stop, by the All-Merciful, stop! Do you not see, O *Pashmuhandis*, that we are drowning? The dams out there are broken, the earth is floating away, the field is under water! Ten feddan of cotton are perishing. Stop, O *Pashmuhandis*. A devil is in your pump and will drown us all.[26]

24 Luke Kemp, Chi Xu, Joanna Depledge, and Timothy M. Lenton, 'Climate Endgame: Exploring Catastrophic Climate Change Scenarios', *PNAS* 119:34, 2022, 6.

25 Ibid., 6–7.

26 Eyth, *Der Kampf*, Chapter 1.

Perhaps the clearest lesson is that the kind of material existence the fossil-powered technologies brought to Egyptian peasants, as much as the entire world, brought with it an undercurrent of justified and legitimate subaltern hesitation about playing with fire. The Faustian deal with the fossil/demonic powers, inaugurated with the Industrial Revolution, perhaps needs to be seen and contemplated in these simple and uncompromising terms.

It was utterly wrong all along. Just look at the terrible price.

Conclusion

How did fossil-fuelled technology find its way to Egypt on its way to global dominance? Let me attempt an answer, based on the times and places this book addressed. It is an answer tied strongly to the trope of crisis, as the manifestation of a nature/society combination mode of analysis. It was crisis that introduced fossil-fuelled technology to Egypt in the nineteenth century, and it was crisis predicated on its purchase and use in expanded commodity production that landed Egypt in decades of European colonialism. The following section elaborates on this statement. The last two sections handle questions of the complicated history of energy transition this book presents, and the implications of this history from Egypt for the planetary crisis of global warming.

Igniting the Steam Engine in Egypt

Starting with Part I, I looked at this central question by positioning fossil-fuelled technology in commodity production in Egypt at the centre of analysis. The analytic of nature/society combinations was deployed with the assumption of the interpenetration and co-evolution of the two. This frame for revisiting the reign of Mehmed Ali (1805–48) showed that he sought and managed to build a predatory state that grew gradually in pursuit of limitless growth in agrarian production for trade.

Out of expansion in global trade in a few monocrops came in the 1820s serious attempts at industrialisation to trade in higher-value mass-produced commodities, and more successful endeavours to build an army and a regional empire to protect his trade and provide it with production inputs. Notably, the inputs in question included coal, to optimise this mass production of commodities through industrialisation. By the 1830s, the oxen-modern-factory-machine hybrid, which was never replaced with secure supplies of coal to ignite the fire of a fossil empire on the Nile, entered domestic, regional, and global crises. The agrarian producers revolted, maimed their own bodies, and fled their fields en masse, depleting the tax base and agrarian production in this process of resistance.

Regionally, the search for coal in Syria floundered, and the British fossil empire stepped in, defeating the Pasha's army of 100,000 soldiers, and imposed free trade by force. When Mehmed Ali tried to evade this imposition that was to put an end to his monopoly on commodities for trade – the alpha and omega of his predatory state – he redistributed the land to his family and entourage, creating, accidentally, the landed-nobility class in the early 1840s. Nature, socially augmented by the necessities of global trade in animal power, hit back through the cattle plague of the early 1840s, when the new-found class suddenly imported thousands of oxen – raised then transported on ships in concentrated populations, heightening the risk of the spread of epizootics across whole batches – to work the newly acquired land and feed the Pasha with commodities to keep the monopolies going in evasion of the British-imposed free trade. The oxen-modern-factory-machine hybrid in the manufactories of the Pasha fell in this disaster, and was almost completely discarded – and it lost its grand purpose after the army was radically downsized – while agrarian production suffered until the plague faded. When fossil-fuelled technologies finally found their way to the farm and factory in Egypt in the early 1850s, it was this new land-holding class that brought it in under conditions of dependency on European fossil capital, financial capital, and within a range of choices that never challenged the growing British fossil empire.

The upwards curve of Mehmed Ali's Pashalik was frustrated, and his budding empire fell with him. What is missing in the history of industrialisation across the first half of nineteenth-century Egypt is precisely the thing that fleshes out this curve and its downfall most clearly:

Mehmed Ali's prospecting for coal, domestically, in Sudan, and across Greater Syria – the fuel that was powering the modern machines of the Industrial Revolution. In the absence of coal (which was sought after by the Pasha to unleash an independent fossil empire from the Nile valley to challenge the British regionally), Egypt's integration in the Capitalocene was never realised in a way that made it one of its champions, those who burned most of the coal that was mined across the nineteenth century. When chimneys of Cornish steam irrigation pumps and sugar factories started sending dense columns of black smoke into the horizon across the valley, they did so to produce and export cotton (its best strains needed many rounds of irrigation) and white sugar (to meet rising expectations on its quality) to international markets, from the fields of the richest land proprietors, including members of the Viceroyal family.

The downwards part of the curve started to build up most seriously with Sa'ed Pasha (1854–63), who imposed a system of cash taxation and the monetisation of the agrarian economy, oversaw the stronger penetration of European financial capital, and witnessed the spread of modern steam-powered cotton-ginning establishments across the Delta by European and Egyptian capital. This was an industrial process that intervened intimately, as we have seen in Chapter 4, in cycles of microcredit, in which the small and middle farmers got entangled while they expanded their cotton cultivation to cater for cash taxation to the state and the global cotton famine caused by the American Civil War at the turn of the 1860s.

Between the 1810s and the early 1860s, the deployment of fossil-fuelled technology in agrarian production in Egypt – always seen and analysed within the flows of the global march of the Capitalocene, led by the British fossil empire – followed a dynamic of expansion through crises. The crisis that opened the doors for feeding Egypt-as-a-commodity-frontier steam-powered technology to produce the commodities in question was the cattle plague and the redistribution of land in the 1840s. The crisis that brought steam technology to the Nile before 1863 – steam-powered cotton gins – was that of the cotton famine in the early 1860s.

At this moment – especially starting in early 1863, when Ismail Pasha ascended the throne – we can see a process coming into sharp relief, through which the latching of steam-powered technology to the mass production of commodities for trade engulfed Egypt, a process of deploying fossil capital with dependence on the financial capital (international loans) that had accumulated earlier in Europe.

In Part II, I challenged the mainstream narrative on Khedive Ismail as a wasteful oriental prince who was deceived by premeditated European designs for colonising Egypt by entangling it in irrational borrowing. Here, I tried to account for how fossil capital tended historically to become increasingly globalised, and how capitalism tended historically to spread into agrarian production frontiers.

Ismail's reign started, in January 1863, as a perfect storm of nature/society combinations that coevolved at an accelerated pace into multiple crises, bringing a sudden leap in the spread of fossil-fuelled technologies to Egypt. The cotton famine raged on at the moment when the ambitious mega-land-proprietor – Ismail – came to power. He quickly consolidated his landholdings, confiscating at will from farmers across the valley and Delta. He suddenly imported thousands of cattle heads to power the sought-after expansion in the Daira Saniya – his own huge latifundium – to meet the demands from the global cotton famine. Unsurprisingly, following the trend of land redistributions that brought vulnerability to epizootics in the 1840s, the cattle plague returned with fury, killing almost all animal power in Egypt, including the Khedive's own.

Over in England, it was also, accidentally or not, the moment when the steam cultivator, which was originally developed to plough heavy and wet English soil, came to prominence. Inspired by his uncle, Prince Halim – one of the richest land proprietors from the 1850s rising in the wake of Mehmed Ali's frustrated empire, who invested heavily in steam cultivation and irrigation – the Khedive imported hundreds of machines and steam engines to consolidate his wealth and power. He created in the process a steam-powered agrarian economy unwitnessed until then anywhere else in the world, not even in England. By the early 1870s, the steam cultivator tilted the balance from animal-human-powered to steam-powered irrigation. It straddled rural Egypt in the form of the steam irrigation pump for growing cotton and sugar cane, bought from British fossil capitalists who were mass-producing steam engines for agrarian applications, and bought through loans from Europe. The sudden increase in fossil-fuelled technology soon unleashed its own socioecological contradictions. By 1865, the searing taxation of the peasants of Middle Egypt in corvée labour to build huge infrastructures to accommodate the steam engines triggered what was termed by a contemporary 'a communist' rebellion that sought to redistribute all wealth, but was soon extinguished by the brute force that visited Qau at the Girga–Assiout line on board Nile steamers.

The same event – a sudden leap in fossil-fuelled technology – spurred innovation in finding alternative fuels, soon found in biomass. Sugar cane refuse, cotton stalks, and other material from the field that usually went back to the soil as fertiliser became the fuel running the engines to produce the very same commodities. We saw how this twist might have inspired more expansion in modern/fully mechanised sugar production, accumulating in the process more debts to European financial capital, bringing colonialism by explicit military force closer. Throughout the remaining years of the nineteenth century, however, coal imports from Britain to Egypt kept increasing, regardless of this seemingly transient shift to biofuels on the field that evidence shows was ongoing until the 1890s.

After covering what might be the essential synthesis of the ignition of the steam-powered machine on the Nile, I asked how fossil-fuelled technology found its way to Egypt on its way to global dominance and once again applied that question to elite land owning intellectuals and state officials and the masses of peasants, those on the ground at the moment the engine gathered steam, and in the decades after. The elite landholding state officials, focusing on Ali Mubarak and Rifa'a al-Tahtawi, fleshed out – respectively – the scientific thermodynamic logic of the steam engine, and an economic system of *industrie* and industrial colonialism (otherwise known as international trade in al-Tahtawi's words) geared towards the generalised production of the commodity across expanding geographies. Those on whose lands the steam engine was ignited, and felt the dark side of its heat, had entertained well into the twentieth century an uneasy relationship to the self-acting machines, full of scepticism and fear, and associated with sacrificial ritual practices. The cruel and absurd industrial accidents that came with the engine might have been the immediate triggers of these views, but the folktales, fables, and stories about fossil-fuelled technology also connoted subtle rationalisation of the moments of the rise of the Capitalocene in the Nile valley, the replacement of the slower naturally cadenced prime movers with the new motive powers. These are also narratives fit for our age, acting as cautionary tales against the state of our present and the future of the mechanised commodity-producing machines of the world, taking the planet relentlessly – by the quest, with matching power, for limitless growth – towards the point of no return.

Based on this history, two issues remain to be addressed, on the transition to steam power for agrarian commodity production in Egypt across the nineteenth century. The first point is about contingency in the transition to steam power, and the second is about the anomaly of running steam engines throughout a few decades in nineteenth-century Egypt on biomass and what this says about assumptions we have on the crucial role of steam engines in igniting the fire of global warming.

Three Peculiarities: The Contingencies of Egypt's Energy Transition

In the absence of socioecological analysis on the history of the burning of fossil fuels, we might tend to envision energy transitions – ahistorically – to be linear, clear-cut, and of an almost evolutionary character. The question of energy transition is framed and thought of frequently as if advocates for renewable energies and relevant policymakers can wilfully transition out of burning fossil fuels by the force of planning and crafting policies for the energy sector.

The history of fossil-fuelled technologies in agrarian production in Egypt presents a few moments of contingency that break with such a view, with the effect of reframing what the transition to fossil fuels was and what the transition out of it might look like. What are these moments and what do they imply regarding the nature of the transition to fossil-fuelled technology in Egypt?

The first moment is that of Mehmed Ali Pasha's search for and failure to find coal mines in the regions he governed and annexed to his short-lived empire. The Pasha managed – without coal – to build an economy that revolved around agrarian commodities traded on international markets. He managed to build a strong army that stormed the region in less than two decades and reached the gates of the Ottoman capital. Mehmed Ali found and mined coal in Syria – which he might have invaded for this very reason – but its quality proved disappointing to all purposes. Britain, the first comer to steam-powered technology whose deployment hinged on access to abundant coal, possessed the power to decisively crush this contender who lacked in comparable coal endowments. This means that when the transition to steam energy was initiated, in its early phase before it reached Egypt, it was a transition highly concentrated in space.

Steam power was controlled and used by Britain – who developed and possessed it in proportions far exceeding other European powers – in geopolitical imperialist conflicts. While the rise of fossil capital and the technologies it fuelled built the Industrial Revolution at home, its sustained use was rendered useful in building a fossil empire across the globe. When fossil-fuelled technologies started to increase radically in Egypt in the later decades after Mehmed Ali's predatory project was defeated, it did so through dependency on Europe, in ways that were always consistent with the interests of the expanding British fossil empire. The timeline of increasing Egyptian coal imports from Britain across the second half of the nineteenth century and into the twentieth is matched by a timeline of increasing dependency on Europe – where British capital featured prominently – then this dependency took a leap with the British occupation of 1882.

The implications of this for Egypt's later moments of transition to fossil-fuelled technologies are that they were shaped by – and took part in shaping – the growing dependency on the core: the British fossil empire and other European powers. In other words, the subsequent contingencies of transition to steam power in Egypt followed from this overarching geopolitical and imperial dynamic where British coal endowments featured prominently.

The second moment of contingency in the transition to steam power is the role of epizootics. The sudden and radical expansion in agrarian production – following sudden and radical land redistributions – which relied on acquiring much more animal power, led to the two epizootic events of the 1840s and the 1860s. In these two episodes, the mass death of animal power massively sped up the arrival of steam power in Egypt, especially in the latter episode. Here, the transition to the steam engine was triggered as an emergency measure to handle a motive-power crisis. This, again, challenges the often-presumed linearity of energy transitions. Let us look more closely at these two events.

Evidence suggests that the rich and powerful landholders who started to install steam engines in the late 1840s and early 1850s acquired their status because of the British pressures at the turn of the 1840s to end Mehmed Ali's monopolies and enforce free trade. The act of redistributing the land, in turn, had a major stake in the spread of the epizootic, since the new landholders rushed to import huge numbers of oxen to make their estates productive, in order to give the Pasha his taxes while making profits from the estates. The epizootic led to the collapse of Mehmed Ali's industrial project with the death of the oxen that ran his machines in the absence of coal. It also led to the deterioration of the irrigation infrastructure, including the summer canals. This is because the Pasha became less able – given the major financial losses in cattle heads and the money spent to replace them – to maintain the works with this aggravated economic crisis. The big Cornish steam engines were imported (partially) to fix this irrigation crisis on the estates of the newly formed rich landholder class by means of mechanically lifting irrigation water from the summer canals. Steam engines were ultimately used to grow and sell agrarian commodities that served as production inputs for fossil capital in Europe, in the wake of the animal power scarcity that lasted for some years in Egypt and perhaps because of fear of the animals' vulnerability to epizootics.

In the early 1860s – in the middle of the global cotton famine – Ismail Pasha once again redistributed the land to create his own huge estates across Egypt. To these lands he imported large numbers of oxen, and thereby came the epizootic crisis that again killed almost all animal power in the country. With this event, the steam-powered irrigation pumps, purchased by loans that deepened dependency on European capital, arrived in unprecedented numbers. Hundreds of engines were imported and installed within months to replace the dead cattle. The steam cultivator – the small 10 hp locomotive engine for ploughing – happened to have been developed commercially in Britain at the turn of the 1860s, and it was soon adapted to irrigate hundreds of thousands of acres under cotton that became very profitable with the cotton famine induced by the American Civil War. Frequent irrigation for the best strains of cotton with highest prices that were processed in the British (and other European) spinning mills was a major factor in the decision to deploy steam power for irrigation, especially with the deepened crisis of deteriorating maintenance of the summer canals that fed the cotton fields.

In both epizootic episodes, we have a new prime mover that was not adopted because of its superior productivity rates over animal prime movers or linear cost-efficiency calculations. The steam engine was ignited on the Nile – in the final analysis – as a provisional fix for links in a long chain of crises coming with the expansion in the production of commodities for their exchange value. Steam power was disseminated at scale mainly because it replaced the dead cattle. No less importantly, because it proved immune to disease that frequently visited these metabolising organisms that powered agriculture whenever the big landholders imported them at scale in order to upscale their agrarian commodities production that found receptive markets in Europe. Another important factor here in the turn to steam is the massive losses landholders and the state sustained when the two epizootics killed their animal power: the millions of pounds lost in both events – money that was equal to or more than average annual state revenue at the time – meant less ability to maintain the extensive irrigation infrastructure, and reluctance to reinvest in animal power. A disease-immune prime mover was needed because enormous energy-intensive mono-cropped plantations proved vulnerable to epizootics in a country that lacked pastures to grow and invest in cattle raising. Hence, agrarian production in Egypt missed the opportunity of having a reserve of locally grown cattle that suffered no congregation in limited space, with better immunity systems. It was a country that tended to import such animals en masse whenever agricultural lands were to be reoriented for such production at scale.

Abrupt expansions in mono-cropping in Egypt tended to bring the ecological collapse of animal power. Next, this collapse of animal power tended to induce massive and sudden transitions to fossil-fuelled prime movers. The energy transition that came with the Industrial Revolution at the core unleashed – in turn – an energy transition at the periphery through a complex socioecological chain of causalities in order to solve crises of ecological and economic nature. What we are seeing here is the metabolism of global capitalism stretching beyond the limits of nature's metabolism across regions, creating, in the process, multiple socioecological crises and subsequent fixes. This is the same kind of dynamic that came with global warming. More on this in the last section.

The implications of this second moment are that Egypt, after 1863, transitioned to steam power without possessing local coal to run the steam engines for agrarian production with economic efficiency. This is the third and last moment I address.

This third moment of contingency is that of the transition to biofuels within the transition to steam power in Egypt. The combustion of bagasse – effectively powering the production processes of the agrarian-commodities economy – shows that even a transition to steam power did not necessarily mean a shift to burning coal. In England, coal mining led to the invention of the steam engine, and the steam engine kept running on coal when it moved to factories. One was intimately tied to the other for reasons that included coal's abundance underneath British soil. In Egypt, the transition to biofuels within a transition to steam power came to fix yet another crisis, that of the expensive bill of coal imports from England. In hindsight, perhaps the transition to steam power during Ismail's early years in power was a mistake, or a half-measure to handle the power crisis from the 1863 epizootic. It was a decision contingent on the sudden collapse of animal power and the need to keep producing cotton for the markets hungry for it because of the American Civil War. As long as the cotton prices remained high, it was economically feasible to import coal to grow cotton. But the situation soon changed with the end of the war in America by the mid-1860s, and biofuels were the provisional solution to fix this next crisis.

The Egyptian two-layered transition – to biofuel combustion to generate steam power – unleashed its own ecological problems, which were different from those of global-scale coal combustion that ultimately landed us in global warming, while following the same trajectory of successive and mounting crises. I will return to this point in the concluding remark that follows, but, for now, let us flesh out an immediate implication. The innovations in energy conjured then discarded along the fossil energy's long path might not necessarily have been benevolent ones. Coal combustion increases carbon particles in the atmosphere. Biofuels – through the ways in which they were produced in nineteenth-century Egypt and across sugar cane lands and sugar mills during earlier centuries across the globe – left the soil exhausted and promulgated contingent innovations that impoverished the landless and small farmers in Egypt. For example, the expansive deployment of pigeon towers in Middle and Upper Egypt around sugar cane plantations that ate the grains grown by peasants for their household consumption, to provide pigeon guano instead of fertilising the soil under sugar cane with the bagasse.

It was a contradiction in terms, leaving us with the very phrase 'fossil-fuelled technology' as an oxymoron, at least until biofuels were discarded altogether and coal and then oil ignited the engines exclusively, at some point in the early twentieth century. But it was not really a contradiction in terms once we look at the bigger picture, where fossil capital proper – represented by the steadily rising coal imports across the nineteenth century in Egypt – kept finding its way to the Nile valley and expanding. Nothing that comes close to a clean and clear energy transition seems to have been in order. The transition to steam power took the form of contingencies coming up successively in waves to fix problems in crises intimately associated with the expanded production of commodities for their exchange value (that is, capitalism) that earlier waves of solutions created. The result is that we are left with a timeline that runs against what one usually expects in a linear story of energy transition at every turn.

Do You Not See That We Keep Drowning?

What may this complicated tapestry of Egypt's nineteenth-century energy transition mean for a global transition out of fossil fuels? The coming transition will not be a linear one, simply because it was never about a transition of energy itself. It was mainly a transition in ways humans metabolise in nature following the logic of capitalist development since capitalism emerged a few centuries ago. Now, since we are still living through capitalism, the next transition is going to be mixed up with contingent events of all sorts that are associated with capitalism. Let us look at two potential kinds of such events.

The first one – extending the line from Egypt's nineteenth-century epizootics – is pandemics, or the combination of viral diseases and capitalism in short. The tendency of expanded and intensified cash crops for global capitalism to bring the cattle epizootic to nineteenth-century Egypt is eerily similar to the tendency of global capitalism at the present to bring humans devastating epidemics and pandemics. There is a pattern here. In the nineteenth century, it was expansion in mono-cropping that unleashed contingencies that escalated between nature and society to the result of bringing the cattle plague catastrophically to the land. According to a study the World Bank published in 2022 – *Putting Pandemics Behind Us: Investing in One Health to Reduce Risks of Emerging Infectious Diseases* – in the twenty-first century, it is capitalist agricultural expansion into the tropics and associated deforestation that contributed a great deal to the origination of many infectious diseases, including Covid-19. The World Bank says that economic activities that encroach on forests and wildlife habitat 'generate cascades of factors that can modify, accelerate, or amplify the dynamics of pathogens and diseases'. It acknowledged that over one third of new infectious diseases that appeared since the 1960s were related to land-use change, while 15 per cent were tied to forests. This is particularly true 'in the tropics', the study adds.

With the 1863 epizootic, Egypt plunged into steam power. What could perhaps happen if the next emergent viral disease – that will mutate from bats or other animals in devastated and fragmented wildlife habitats – pushes humanity towards far steeper ranges of morbidity and mortality than Covid-19 did? How will such an event fit into the fabric of multiple contingent processes and dynamics towards inducing the next big energy transition?

A second potential kind of event – again learning from Egypt's history with energy transition – is one related to how a global transition out of fossil fuels might unfold through geopolitical and neo-imperialist conflicts, and thus how the choice of fuels in the future might take unforeseen turns. Might the Russia–Ukraine conflict – or Israel's genocidal war on Gaza, or another massive act of violence in the near future – lead to unforeseen decisions and actions towards renewables? Might global food prices sustain their upwards trend through socioecological aggravating events and as such render certain energy systems – for better or worse – more feasible than others? These are the kind of questions that handling the non-linearity of energy history seriously might bring. Perhaps the salient point regarding this line of inquiry is that of the capitalist metabolism that will likely keep rendering emergent energy modes unsustainable and damaging to humans and nature alike, and keep weighing humanity down as it struggles against drowning in global warming and the known and unknown events it is going to bring. Let us conclude with this point.

The capitalist metabolism in nature on the global level brought the notion of the Capitalocene: the collision of the mass production of commodities – through burning fossil fuels – with planetary limits. Capitalism became a geological force, at least metaphorically, once this happened. Capitalism becomes the Capitalocene by virtue of its inherent nature, that of ploughing forward to resolve its crises by constantly jumping into new crises in order to keep reproducing capital and expand across space and inherently limited natural endowments.

The logic of the exploitation of humans and nature is what kept capitalism burning once it was ignited. It is global warming that inspired the coining of the term 'Capitalocene', but capitalism's deeper current, the very stuff of which it is constituted, is exploitation of all sorts regardless of limits. In its conjunction with global warming, it is misery suffered by humans and nature hitting existentially uncrossable limits. Its history is as much about all sorts of exploitation and extractivism as it is about global warming. The socio-economic and political history of nineteenth-century Egypt with fossil-fuelled technology, conjured through then leading up to socio-economic and political disasters, is a powerful impact of capitalism – on a local scale – that predated capitalism's planetary impact of global warming and formed part of its history.

Following from this depiction of capitalism, discontinuing the burning of fossil fuels – if it is stopped by some miracle tomorrow morning – is not enough to start turning the tide of global warming. Fossil energy is, ultimately, embedded in global networks of exploitation of all beings and things; it is embedded in capitalism's first and second contradictions, those of subsuming labour and nature. As such, dismantling fossil energy should not be sought after as a transition to clean energy. It should be, instead, a passage out of capitalism, where clean energy becomes embedded in a mode of just reciprocal existence for all. To become linked to a mode of living, not a mode for the relentless accumulation of profit. It is the Capitalocene, not a geological epoch of fossil fuels. This is precisely because capitalism (the boundless and limitless production of the commodity) is the prime mover that unleashes all sorts of crises and disasters, including, ultimately, global warming. The diabolical problem we have now with global warming is the co-evolution of all sorts of socio-ecological crises together, until we will, very likely, hit a sudden and impenetrable wall, some major human extinction event that is brewing at this very moment. Fossil capital has been, since the nineteenth century, the ferocious mechanism and process of labour and nature exploitation. This unrelenting exploitation needs to be addressed. If it carries on after we magically strike down fossil capital somehow, capitalism will come up with other fixes with other devastating implications for the planet.

The dynamic through which biofuels in nineteenth-century Egypt were deployed instead of coal in steam engines is instructive. Biofuel burning – let us assume – was less polluting to the atmosphere in comparison to burning coal to produce cotton and refined sugar. If we imagine that this production was progressing at a rate just for the subsistence of the producers, its ecological damage could, arguably, have stayed within acceptable limits. An acre of sugar cane would consume bagasse from the harvest of the previous year to produce enough steam power to irrigate the present year's crop, and, at subsistence scale, it could be relatively easy to come up with other natural fertilisers for the soil without necessarily disturbing socioecological balance. Or, alternatively, it could be easy to rotate the crop more, giving the soil enough time to regenerate.

Perhaps, to start with, the farmer would have chosen to grow another less energy-consuming crop with comparable utility as food if he was not growing crops for their exchange value. But a major-scale production process of sugar through powerful sugar refineries is another matter. In such a situation – namely, what really happened – we have cultivated space running into thousands of acres, and every single shred of bagasse, straw, or other combustible organic material (biofuels) is summoned to the furnaces. The average big farmer here, the Khedive, depletes the soil thoroughly, and orders the building of pigeon towers that eat the grains his peasants grow for their sustenance in order to fertilise the sugar cane fields. We end up with impoverished hungry masses and, perhaps, famine. This latter scenario is agrarian production for a capitalist global market. A renewable fuel of sorts is used, but, nevertheless, catastrophe and misery strike.

Alternatively, let us imagine – again for the sake of the argument – that, for some reason, nineteenth-century Britain had awakened to the lethal dangers of coal burning and decided to clone Egypt's experiment with biofuels on an imperial scale. One could imagine a massacre of trees and wildlife habitats across the Amazon rainforest to consume the wood to generate steam power, then to cultivate some highly combustible plant matter at a huge scale because capitalist production demands that we keep burning and growing factories to expand more and more in producing commodities of all sorts. Global warming through fossil-fuel burning leaves through the door, and global warming because of a deforested Amazonia enters from the window.

A more structural moment for examining what capitalism did is how regardless of the spread of biofuels across the 1870s and 1880s, and then well into the twentieth century, Egypt kept importing increasing quantities of coal in a chilling upwards trend that just kept going regardless of the collapse of the sugar factories, state bankruptcy, and military invasion. While biofuels were possible to use in the field and at the sugar factory, their bigger mass versus energy units generated made biofuels more challenging, at least logistically, to handle on board steamers and trains. The extensive fabric of trade and transportation logistics for consuming the green cotton, that is, cotton with a lower carbon footprint, in fossil-powered spinning mills in Britain and elsewhere made it structurally necessary to burn more coal everywhere else beyond the fields.

This was, ultimately, the upward trend of the development of capitalism in Egypt, marching onwards because capitalism always finds ways to expand and reproduce itself as a general logic of limitless growth within limited nature.

Let us contemplate one last summary of what these scenarios are telling us: the very logic of limitless growth for the production of the commodity within a finite planet will always lead to scenarios of socio-ecological collapse. The salient point with capitalism, which turned, eventually, into the Capitalocene, is that, in a *fuite en avant*, it keeps leaping out of one crisis only to land in the next one, and as long as this dynamic forms the marrow of its bones, we all will continue to drown.

Index